Electrical Principles and Practices

fourth edition

Workbook

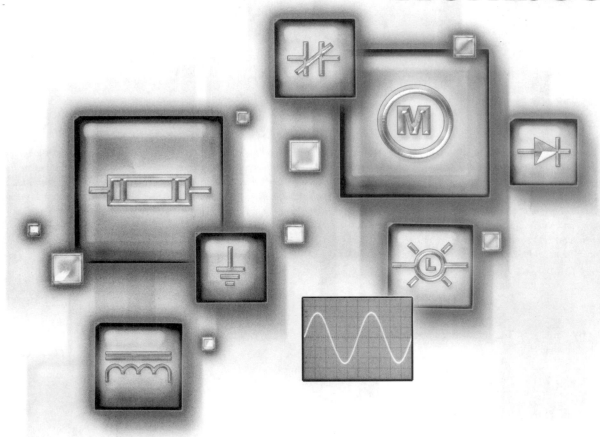

Electrical Principles and Practices Workbook contains procedures commonly practiced in industry and the trade. Specific procedures vary with each task and must be performed by a qualified person. For maximum safety, always refer to specific manufacturer recommendations, insurance regulations, specific job site and plant procedures, applicable federal, state, and local regulations, and any authority having jurisdiction. The material contained is intended to be an educational resource for the user. American Technical Publishers, Inc. assumes no responsibility or liability in connection with this material or its use by any individual or organization.

© 2013 by American Technical Publishers, Inc.
All rights reserved

4 5 6 7 8 9 – 13 – 9 8 7 6

Printed in the United States of America

ISBN 978-0-8269-1812-3

Contents

1 Electricity Principles

Review Questions **1**
Worksheets **3**
   1-1   Elements.. 3
   1-2   Compounds.. 4
   1-3   Chemical Formulas 5
   1-4   Electron Flow... 6
Activities **7**

2 Basic Quantities

Review Questions **9**
Worksheets **13**
   2-1   Energy... 13
   2-2   AC Voltage Values.................................. 14
   2-3   Voltage Waveforms................................ 15
   2-4   Electrical Circuits 16
   2-5   Temperature Conversions...................... 17
   2-6   Electromagnetic Spectrum...................... 18
Activites **19**

3 Ohm's Law and the Power Formula

Review Questions **21**
Worksheets **25**
   3-1   Determining Current, Resistance,
         Voltage, and Power 25
   3-2   Determining Circuit Current 26
   3-3   Applying Ohm's Law
         and the Power Formula 27
   3-4   Determining Conductor Current.............. 28
Activities **29**

4 Safety

Review Questions **31**
Worksheets **35**
   4-1   Fire Extinguisher Classes 35
   4-2   Lockout/Tagout 36
   4-3   Enclosures 37
   4-4   Hazardous Locations............................. 38
Activities **39**

5 Math Principles

Review Questions **41**
Worksheets **45**
   5-1   Conversions... 45
   5-2   Graphs 46
   5-3   Fractions.. 47
   5-4   Percentages 48
   5-5   Prefixes 49
   5-6   Metric System Conversions 50
Activities **51**

6 Math Applications

Review Questions **53**
Worksheets **57**
   6-1   Heating Water and Oil 57
   6-2   Heating Air and Electric Heater Output 58
   6-3   Conductor Maximum Length Calculation........ 59
   6-4   Motor Required Horsepower
         and Pulley Size 60
   6-5   Object Activating Speed
         and Air Resistance........................... 61
   6-6   Blower Sizing 62
Activities **63**

7 Numbering Systems and Codes

Review Questions **65**
Worksheets **67**
   7-1   Binary/Decimal and
         Octal/Decimal Conversion.............. 67
   7-2   Hexadecimal/Decimal Conversion 68
   7-3   Resistor Color Codes 69
   7-4   Capacitor Color Codes 70
Activities **71**

8 Meter Abbreviations and Displays

Review Questions **73**
Worksheets **75**
   8-1   Reading Analog Displays......................... 75
   8-2   Reading Digital Displays/Reading
         Meter Voltages.............................. 76
Activities **77**

9 Taking Standard Measurements

Review Questions **79**
Worksheets **83**
    9-1   AC Voltage Measurements 83
    9-2   DC Voltage Measurements 84
    9-3   In-Line AC Current Measurements.................. 85
    9-4   Clamp-On AC/DC Current Measurements 86
    9-5   Resistance Measurements.............................. 87
    9-6   Oscilloscope/Scopemeter Measurements 88
Activities **89**

10 Symbols and Printreading

Review Questions **91**
Worksheets **95**
    10-1 Abbreviations.. 95
    10-2 Acronyms.. 96
    10-3 Electrical Symbols...................................... 97
    10-4 Electronic Symbols..................................... 98
    10-5 Fluid Power Symbols.................................. 99
    10-6 Component Usage...................................... 100
Activities **101**

11 Circuit Conductors, Connections, and Protection

Review Questions **105**
Worksheets **109**
    11-1 Conductor and Conduit Identification 109
    11-2 Fuses and Circuit Breakers........................... 110
Activities **111**

12 Series Circuits

Review Questions **113**
Worksheets **117**
    12-1 Resistance in Series Circuits 117
    12-2 Voltage in Series Circuits 118
    12-3 Current in Series Circuits............................. 119
    12-4 Power in Series Circuits 120
    12-5 Capacitors and Inductors in Series Circuits 121
    12-6 Batteries and Solar Cells in Series Circuits...... 122
Activities **123**

13 Parallel Circuits

Review Questions **125**
Worksheets **129**
    13-1 Resistance in Parallel Circuits 129
    13-2 Current in Parallel Circuits 130
    13-3 Voltage in Parallel Circuits 131
    13-4 Power in Parallel Circuits 132
    13-5 Capacitors and Inductors in Parallel Circuits.................................... 133
    13-6 Batteries and Solar Cells in Parallel Circuits.................................... 134
Activities **135**

14 Series/Parallel Circuits

Review Questions **137**
Worksheets **141**
    14-1 Resistance in Series/Parallel Circuits 141
    14-2 Current in Series/Parallel Circuits 142
    14-3 Voltage in Series/Parallel Circuits 143
    14-4 Power in Series/Parallel Circuits 144
    14-5 Power, Resistance, and Current in Series/Parallel Circuits................................ 145
    14-6 Capacitors and Inductors in Series/Parallel Circuits................................ 146
Activities **147**

15 Transformers and Smart Grid Technology

Review Questions **149**
Worksheets **151**
    15-1 Transformer Operation................................ 151
    15-2 Transformer Overloading and Temperature Compensation 152
    15-3 Sizing 1ϕ Transformers 153
    15-4 Sizing 3ϕ Transformers 154
    15-5 Transformer Current Draw........................... 155
    15-6 Transformer Connections 156
Activities **157**

16 Electric Motors

Review Questions **159**
Worksheets **163**
    16-1 Split-Phase Motors..................................... 163
    16-2 Capacitor Motors 164
    16-3 Three-Phase Wye-Connected Motors 165
    16-4 Three-Phase High Voltage Delta-Connected Motors 166
    16-5 Three-Phase Low Voltage Delta-Connected Motors 167
    16-6 DC Motors 168
Activities **169**

17 Resistance, Inductance, and Capacitance

Review Questions **171**
Worksheets **173**
    17-1 Inductive Reactance.................................... 173
    17-2 Series Inductive Reactance 174
    17-3 Parallel Inductive Reactance 175
    17-4 Capacitive Reactance 176
    17-5 Series Capacitive Reactance......................... 177
    17-6 Parallel Capacitive Reactance....................... 178
Activities **179**

18 Circuit Requirements

Review Questions 181
Worksheets 183
   18-1 Heating Element Voltage Variations 183
   18-2 Manufacturer's Lamp
       Characteristics Chart Use 184
   18-3 Mechanical Relay Interface Application 185
   18-4 Solid-State Relay Interface Application 186
   18-5 Contactor Interface Application 187
   18-6 Motor Starter Interface Application 188
Activities 189

19 Residential Circuits

Review Questions 191
Worksheets 193
   19-1 Two-Way and Three-Way
       Switch Lamp Control 193
   19-2 Three Location Lamp Control 194
   19-3 Combination Lamp/Receptacle
       Circuit Wiring 195
   19-4 Two Location Receptacle Control 196
   19-5 Lamp/Receptacle Switch Control 197
   19-6 Combination Circuit Wiring 198
Activities 199

20 Commercial Circuits

Review Questions 201
Worksheets 205
   20-1 120/240 V, 1φ, 3-Wire Service Wiring 205
   20-2 120/208 V, 3φ, 4-Wire Service Wiring 206
   20-3 120/240 V, 3φ, 4-Wire Service Wiring 207
   20-4 277/480 V, 3φ, 4-Wire Service Wiring 208
   20-5 Conductor Color Coding 209
   20-6 Lamp Wiring 210
Activities 211

21 Industrial Circuits

Review Questions 213
Worksheets 217
   21-1 Reference Number Addition 217
   21-2 Control Switch Addition 218
   21-3 Sequence Control Circuit 219
   21-4 Indicator Lamp Addition 220
   21-5 Interlocking Circuits 221
   21-6 Manual Control Switch Addition 222
Activities 223

22 Fluid Power Circuits

Review Questions 227
Worksheets 229
   22-1 Product Flattening Application 229
   22-2 Part Eject Application 230
   22-3 Backhoe Application 231
   22-4 Part Feed Application 232
   22-5 Pick-and-Place
       Robot Application 233
   22-6 Part Rotating Application 234
Activities 235

23 Audio Systems

Review Questions 239
Worksheets 243
   23-1 Basic Speaker Connections 243
   23-2 Combination Speaker Connections 244
   23-3 Speaker Wire Sizing 245
   23-4 Basic Meter Testing 246
Activities 247

24 Electronic Control Devices

Review Questions 249
Worksheets 251
   24-1 Diode Current Flow 251
   24-2 NPN Transistor Connections 252
   24-3 PNP Transistor Connections 253
   24-4 SCR Connections 254
   24-5 Triac Connections 255
   24-6 Electronic Device Connections 256
Activities 257

25 Digital Electronic Circuits

Review Questions 261
Worksheets 263
   25-1 Truth Tables 263
   25-2 Electrical to Digital Circuit Conversion 264
   25-3 Digital to Electrical Circuit Conversion 265
   25-4 Digital Circuit Application 266
   25-5 Truth Table Application 267
   25-6 Common Control Circuits 268
Activities 269

Appendix 273

CD-ROM Contents

- Using this Interactive CD-ROM
- Electrical Power Data Sheets
- TECO SG2 Client Simulation Software
- Color Rendering Simulation
- Virtual Meters
- Grounding Application Videos

- Lamp Data Sheets
- Compactor Print
- Ampacities of Insulated Conductors
- Audio Frequencies Interactive Simulation
- ATPeResources.com

Electrical Principles and Practices Workbook is designed to reinforce the concepts presented in *Electrical Principles and Practices*. Each chapter in the workbook covers information from the corresponding chapter in the textbook. The appropriate textbook pages should be carefully reviewed before completing the review questions, worksheets, and activities in the workbook. When studying the textbook, particular attention should be paid to illustrations, examples, and italicized terms.

The review questions consist of true-false, multiple choice, completion, identification, and calculation questions based on the text and illustrations in the corresponding chapter in the textbook. The worksheets provide an opportunity to apply the concepts in each chapter of the textbook to practical electrical circuit design and troubleshooting problems.

This edition also includes activities based on the concepts introduced in the textbook, information provided on the workbook CD-ROM, and the components kit. The activities are designed to engage the learner and present real-world procedures and applications. The CD-ROM contains information used to answer activity questions. Also available from American Tech is a components kit that provides an opportunity for learners to develop basic electrical circuits and to study circuit design, operation, and troubleshooting. The components required to work the activities includes the following:

- 9 VDC power source
- 10 test leads
- 500 Ω potentiometer
- Resistors:
  20 Ω, 75 Ω, 100 Ω, 150 Ω, 200 Ω,
  510 Ω,1 kΩ, 5.6 kΩ, 10 kΩ, 20 kΩ
- Diode
- Red lamp
- Yellow lamp

- Green lamp
- Relay
- Rocker switch
- SPDT limit switch
- Normally open momentary pushbutton
- Locking pushbutton
- Buzzer
- Motor

Information presented in *Electrical Principles and Practices Workbook* addresses common electrical principles and applications. Additional educational materials related to this and other topics are available from ATP. To obtain information about ATP products, visit the ATP web site at www.go2atp.com.

The Publisher

Review Questions test for comprehension of content covered in the corresponding chapter of the textbook.

Activities are designed to engage the learner and introduce real-world procedures and applications.

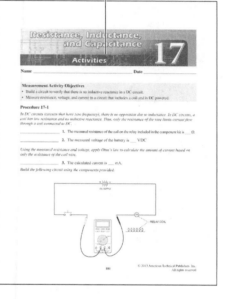

Worksheets apply the concepts presented in the textbook to practical electrical circuit design and troubleshooting problems.

The available Components Kit consists of electrical devices that provide an opportunity for learners to develop basic electrical circuits and to study circuit design, operation, and troubleshooting.

The CD-ROM offers Electrical Power Data Sheets, TECO SG2 Client Simulation Software, a Color Rendering Simulation, Virtual Meters, Grounding Application Videos, Lamp Data Sheets, a Compactor Print, an Ampacities of Insulated Conductors chart, an Audio Frequencies Interactive Simulation, and a link to ATPeResources.com.

Name _____ Date _____

True-False

T   F   **1.** Spalling is the formation of film on the contact surface.

T   F   **2.** A positive charge is an electrical charge produced when there are fewer electrons than normal.

T   F   **3.** An atom is the smallest particle that an element can be reduced to and still maintain the properties of that element.

T   F   **4.** An insulator is material that has very little resistance and permits electrons to move through it easily.

T   F   **5.** Static energy is the stored energy a body has due to its position, chemical state, or condition.

T   F   **6.** A helium atom has the fewest protons of all atoms, has the least amount of mass, and is assigned the atomic number of one.

T   F   **7.** Tungsten contacts are used in high-voltage applications because tungsten has a high melting temperature and is less affected by arcing.

T   F   **8.** There are only 118 elements, but these elements can be combined in many different ways to form millions of compounds.

Multiple Choice

_____   **1.** A(n) ___ is a substance that cannot be chemically broken down and contains atoms of only one variety.
　　　　A. base
　　　　B. element
　　　　C. compound
　　　　D. atom

_____   **2.** A(n) ___ is a combination of the atoms of two or more elements.
　　　　A. substance
　　　　B. atom
　　　　C. element
　　　　D. compound

_____ **3.** ___ is anything that has mass and occupies space.
       A. Matter
       B. Light
       C. Energy
       D. Electricity

_____ **4.** A ___ is a state of matter that has a definite volume and shape.
       A. gas
       B. liquid
       C. solid
       D. vapor

_____ **5.** Most contacts include ___ because it has the highest electrical conductive property of all materials.
       A. iron
       B. silver
       C. aluminum
       D. copper

Completion

_____ **1.** A(n) ___ is a material that has a very high resistance and resists the flow of electrons.

_____ **2.** ___ electricity is an electrical charge at rest.

_____ **3.** A(n) ___ is a negatively charged particle in an atom.

_____ **4.** ___ current flow is from negative to positive.

_____ **5.** A(n) ___ is the smallest particle that a compound can be reduced to and still possess the chemical properties of the original compound.

_____ **6.** A(n) ___ is an electronic device that has electrical conductivity between that of a conductor and that of an insulator.

_____ **7.** A(n) ___ charge is an electrical charge produced when there are more electrons than normal.

_____ **8.** A(n) ___ is a particle with a positive electrical charge of 1 unit that exists in the nucleus of an atom.

_____ **9.** ___ current flow is from positive to negative.

_____ **10.** A(n) ___ is a neutral particle, with a mass approximately the same as a proton, that exists in the nucleus of an atom.

_____ **11.** A(n) ___ shell is the outermost shell of an atom and contains the electrons that form new compounds.

_____ **12.** A(n) ___ fuel is formed by plant and animal remains taken from the ground.

Electricity Principles

Worksheet 1-1

Name _____ Date _____

Elements

Using the Chemical Elements table in the appendix, write the chemical symbol for each element.

_____ **1.** Oxygen = ___

_____ **2.** Carbon = ___

_____ **3.** Hydrogen = ___

_____ **4.** Nitrogen = ___

_____ **5.** Phosphorus = ___

_____ **6.** Sulfur = ___

_____ **7.** Calcium = ___

_____ **8.** Chlorine = ___

_____ **9.** Magnesium = ___

_____ **10.** Potassium = ___

_____ **11.** Aluminum = ___

_____ **12.** Copper = ___

_____ **13.** Gold = ___

_____ **14.** Iron = ___

_____ **15.** Lead = ___

_____ **16.** Mercury = ___

_____ **17.** Nickel = ___

_____ **18.** Silver = ___

_____ **19.** Tin = ___

_____ **20.** Tungsten = ___

Electricity Principles

Worksheet 1-2

Name _____ Date _____

Compounds

Write the element name for each part of the compound.

1.

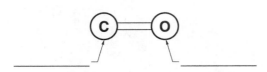

CARBON MONOXIDE

2.

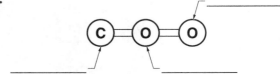

CARBON DIOXIDE

3.

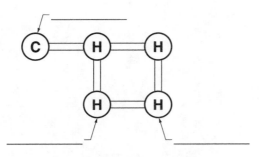

METHANE

4.

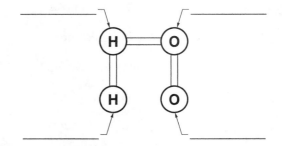

HYDROGEN PEROXIDE

5.

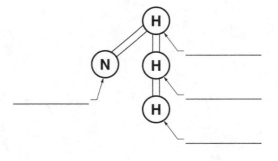

AMMONIA

6.

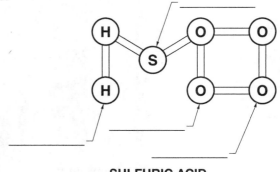

SULFURIC ACID

Name _____ Date _____

Chemical Formulas

Write the chemical formula for each compound or chemical.

_____ **1.** Methane

                formula = ___

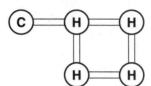

_____ **2.** Hydrogen peroxide

                formula = ___

_____ **3.** Ammonia

                formula = ___

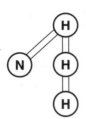

_____ **4.** Sulfuric acid

                formula = ___

_____ **5.** Nitrogen

                formula = ___

_____ **6.** Ethylene

                formula = ___

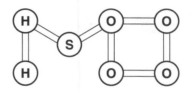

_____ **7.** Lime

                formula = ___

_____ **8.** Phosphoric acid

                formula = ___

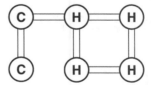

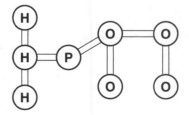

Name _____ **Date** _____

Electron Flow

1. Draw arrows to show the current flow through the low beam circuit using the conventional current flow theory.

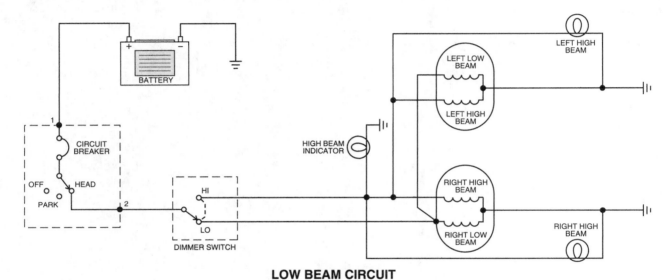

LOW BEAM CIRCUIT

2. Draw arrows to show the current flow through the high beam circuit using the conventional current flow theory.

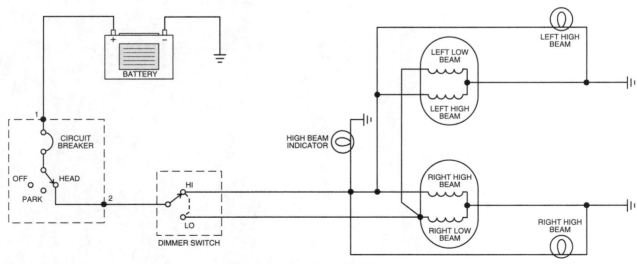

HIGH BEAM CIRCUIT

Electricity Principles

Activities

1

Name _____ Date _____

Element Usage Activity Objectives

- Identify elements used in lamp bulbs that require special handling when disposing of the bulbs.
- Recognize how new lamps are better for the environment than old lamps because they use reduced amounts of hazardous elements.

Procedure 1-1

Lamp bulbs include elements that improve light output, efficiency, bulb life, and create different light-rendering characteristics. Although some elements added to lamp bulbs are not dangerous, other elements that may be added are dangerous. The elements included in a bulb must be known so that proper care can be taken when handling and disposing of the bulb. Click on the Lamp Data Sheets button on the CD-ROM. Next, click on the High-Pressure Sodium Lamp button and review the warning, cautions, and operating instructions section.

1. What two chemical elements are contained in the bulb that require special caution when disposing of the bulb?

_____ 2. What are the chemical element symbols for these two elements?

Click on the Lamp Data Sheets button on the CD-ROM. Next, click on the Energy Saver Dimmable Lamp button.

_____ 3. The energy cost savings for a 20 W dimmable lamp over a 75 W standard incandescent lamp operated for 4000 hours is $___.

Electrical Power Generation Activity Objectives

- Based on the table, determine the increase or decrease in the amount of generated power in the United States.
- Determine which energy sources have had the greatest increase and decrease in production over the last year.
- Determine which state generates the most power and which state generates the least power.

Procedure 1-2

The United States government tracks the amount of electrical power generated in the United States. The information provided is based on fuel type such as coal, hydroelectric, solar, etc. Click on the Electrical Power Data Sheets button on the CD-ROM. Next, click on the Total Electric Power Industry Summary Statistics button and answer the following questions.

_____ 1. Using the total for all sectors, how much more or less net electric power (in thousand megawatthours) was generated in August 2012 than in August 2011 for all energy sources?

_____ 2. Of all the energy sources, which type produced the most electric power in 2012?

_____ 3. Has the amount of electric power generated from solar energy increased or decreased from 2011 to 2012?

_____ 4. Using the total for all sectors, which energy source increased the greatest amount in percentage of electric power generated from 2011 to 2012?

_____ 5. Using the total for all sectors, which energy source decreased the greatest amount in percentage of electric power generated from 2011 to 2012?

Click on the Electrical Power Data Sheets button on the CD-ROM. Next, click on the Net Generation by State, by Sector button and answer the following questions.

_____ 6. In 2012 through August, which state generated the greatest amount of electricity?

_____ 7. How many thousand megawatthours of electricity were generated in this state?

_____ 8. In 2012 through August, which state generated the least amount of electricity?

_____ 9. How many thousand megawatthours of electricity were generated in this state?

_____ 10. Which state had the greatest percentage increase in electricity generated from 2011 to 2012?

_____ 11. Which state had the greatest percentage decrease in electricity generated from 2011 to 2012?

Name _____ **Date** _____

True-False

T   F   **1.** Polarity is the positive (+) or negative (−) state of an object.

T   F   **2.** A capacitor is a device that converts AC voltage to DC voltage by allowing the voltage and current to move in only one direction.

T   F   **3.** Apparent power is the actual power used in an electrical circuit.

T   F   **4.** An insulator is a material that has very little resistance and permits electrons to move through it easily.

T   F   **5.** Thermal conductivity is the property of a material to conduct heat in the form of thermal energy.

T   F   **6.** Electron current flow is current flow from negative to positive.

T   F   **7.** Illumination is the effect that occurs when light falls on a surface.

T   F   **8.** In phase is the state when voltage and current reach their maximum amplitude and zero level simultaneously.

Multiple Choice

_____ **1.** The ___ value of a sine wave is the maximum value of either the positive or negative alternation.
    A. peak
    B. rms
    C. average
    D. peak-to-peak

_____ **2.** A(n) ___ circuit is a circuit in which current leads voltage.
    A. resistive
    B. inductive
    C. capacitive
    D. forward-biased

9

_____ **3.** ___ is the capacity to do work.
A. Resistance
B. Force
C. Power
D. Energy

_____ **4.** Common insulators include rubber, plastic, and ___.
A. air
B. glass
C. paper
D. all of the above

_____ **5.** The color of light is determined by its ___.
A. temperature
B. wavelength
C. voltage level
D. power

Completion

_____ **1.** ___ energy is stored energy a body has due to its position, chemical state, or condition.

_____ **2.** ___ energy is the energy of motion.

_____ **3.** ___ is the amount of electrical pressure in a circuit.

_____ **4.** ___ voltage is voltage that flows in one direction only.

_____ **5.** A(n) ___ is one complete wave of alternating voltage.

_____ **6.** ___ is the amount of electrons flowing through an electrical circuit.

_____ **7.** ___ current flow is current flow from positive to negative.

_____ **8.** ___ voltage is voltage that reverses its direction of flow at regular intervals.

_____ **9.** ___ is the rate of doing work or using energy.

_____ **10.** ___ is the property of a circuit that causes it to oppose a change in current due to energy stored in a magnetic field.

_____ **11.** ___ is the ability to store energy in the form of an electrical charge.

_____ **12.** ___ is the portion of the electromagnetic spectrum that produces radiant energy.

_____ **13.** A(n) ___ is an output device that converts electrical energy into light.

_____ **14.** The ___ value of a sine wave is the mathematical mean of all instantaneous voltage values in the sine wave.

_____ **15.** ___ is the state when voltage and current in a circuit do not reach their maximum amplitude and zero level simultaneously.

_____ **16.** ___ is the opposition to the flow of current.

_____ **17.** The ___ value is the value measured from the maximum positive alternation to the maximum negative alternation of a sine wave.

_____ **18.** A(n) ___ is the number of electrons passing a given point in one second.

_____ **19.** ___ is the measurement of the intensity of heat.

_____ **20.** A(n) ___ is the unit used to measure the total amount of light produced by a light source.

_____ **21.** ___ is the ratio of true power used in an AC circuit to apparent power delivered to the circuit.

_____ **22.** ___ light is the portion of the electromagnetic spectrum to which the human eye responds.

Sine Wave Values

_____ **1.** Peak value

_____ **2.** Peak-to-peak value

_____ **3.** rms value

_____ **4.** Average value

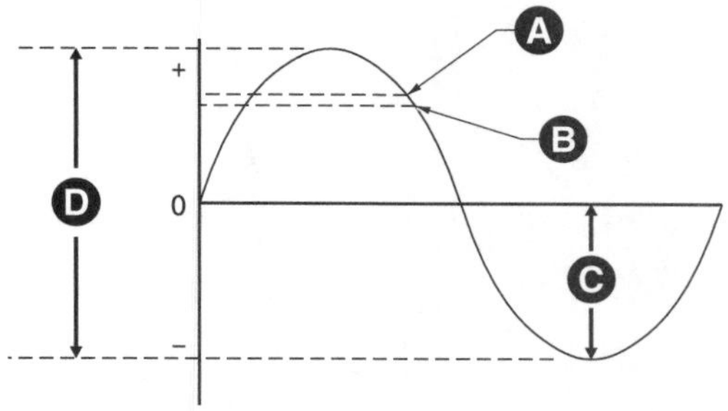

Name _____ **Date** _____

Energy

Determine if each energy source is potential or kinetic.

_____ **1.** Crude oil is ___ energy.

_____ **2.** Rotating blades are ___ energy.

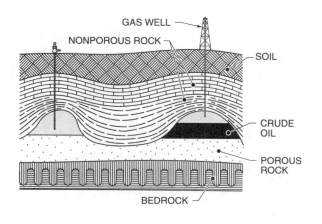

GAS WELL

NONPOROUS ROCK

SOIL

CRUDE OIL

POROUS ROCK

BEDROCK

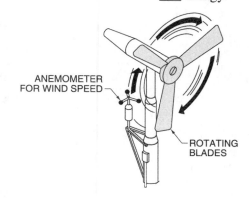

ANEMOMETER FOR WIND SPEED

ROTATING BLADES

_____ **3.** A rotating motor is ___ energy.

_____ **5.** A cylinder ejecting a part is ___ energy.

_____ **4.** Stored compressed air is ___ energy.

_____ **6.** A carton of candles is ___ energy.

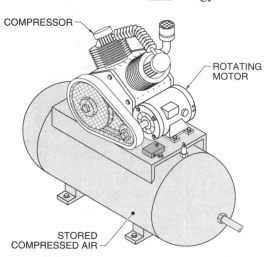

COMPRESSOR

ROTATING MOTOR

STORED COMPRESSED AIR

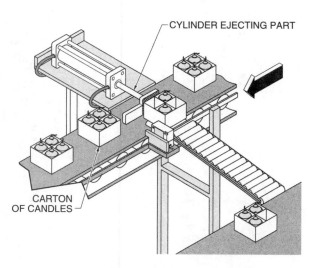

CYLINDER EJECTING PART

CARTON OF CANDLES

Name _____ **Date** _____

AC Voltage Values

_____ **1.** $80 V_{max} = \underline{\quad} V_{p\text{-}p}$

_____ **2.** $80 V_{max} = \underline{\quad} V_{rms}$

_____ **3.** $680 V_{max} = \underline{\quad} V_{rms}$

_____ **4.** $680 V_{max} = \underline{\quad} V_{avg}$

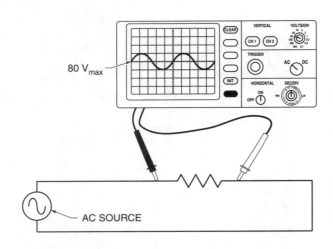

80 V_{max}

AC SOURCE

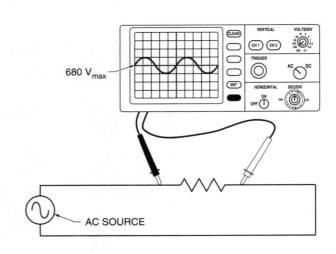

680 V_{max}

AC SOURCE

_____ **5.** $340 V_{p\text{-}p} = \underline{\quad} V_{avg}$

_____ **6.** $340 V_{p\text{-}p} = \underline{\quad} V_{rms}$

_____ **7.** $17 V_{max} = \underline{\quad} V_{rms}$

_____ **8.** $17 V_{max} = \underline{\quad} V_{p\text{-}p}$

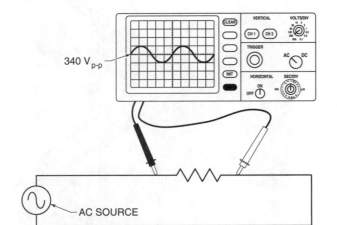

340 $V_{p\text{-}p}$

AC SOURCE

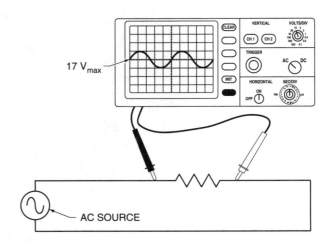

17 V_{max}

AC SOURCE

Name _____ Date _____

Voltage Waveforms

Match the waveform to the corresponding circuit part.

_____ **1.**

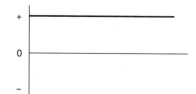

_____ **2.**

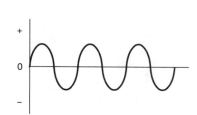

_____ **3.**

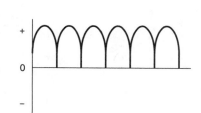

_____ **4.**

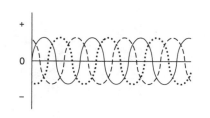

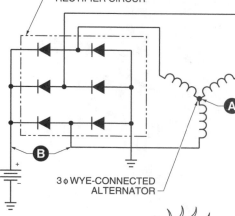

FULL-WAVE BRIDGE RECTIFIER CIRCUIT

3 φ WYE-CONNECTED ALTERNATOR

A

B

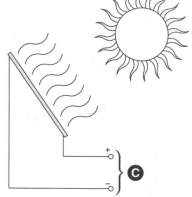

C

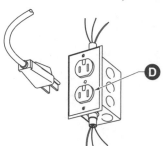

D

Name _____ Date _____

Electrical Circuits

Determine if each load is resistive, inductive, or capacitive.

_____ **1.** Load is ___.

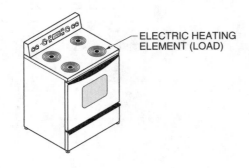

ELECTRIC HEATING
ELEMENT (LOAD)

_____ **2.** Load is ___.

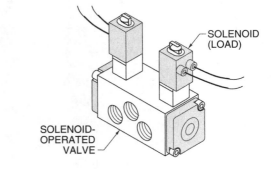

SOLENOID
(LOAD)

SOLENOID-
OPERATED
VALVE

_____ **3.** Load is ___.

MOTOR (LOAD)

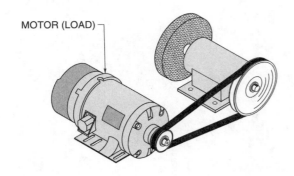

_____ **4.** Load is ___.

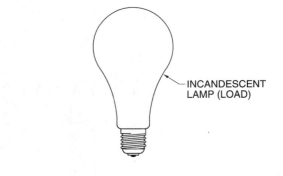

INCANDESCENT
LAMP (LOAD)

_____ **5.** Load is ___.

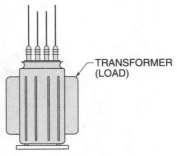

TRANSFORMER
(LOAD)

_____ **6.** Load is ___.

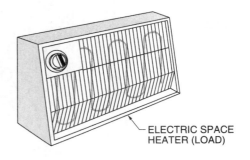

ELECTRIC SPACE
HEATER (LOAD)

Name _____ Date _____

Temperature Conversions

_____ **1.** 180°F = ___°C

_____ **2.** 70°C = ___°F

_____ **3.** 120°F = ___°C

_____ **4.** 35°C = ___°F

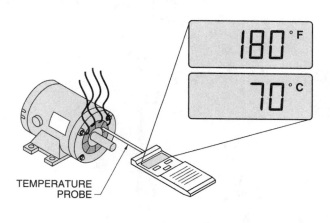

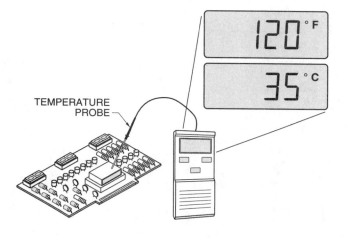

_____ **5.** 630°F = ___°C

_____ **6.** 212°C = ___°F

_____ **7.** 154°F = ___°C

_____ **8.** 62°C = ___°F

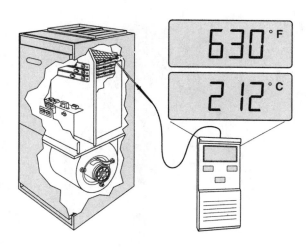

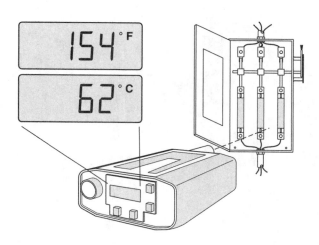

Name _____ Date _____

Electromagnetic Spectrum

Match the wavelength to the correct device.

_____ **1.** 10^{-5} mm

_____ **2.** 400 nm – 700 nm

_____ **3.** 0.1 mm

_____ **4.** 100 m

_____ **5.** 10^6 m

INFRARED RADIANT ENERGY WAVES

HEATING SURFACE

A

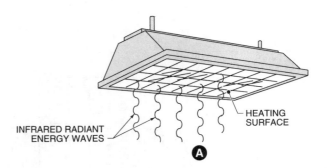

INCOMING RADIO WAVES

RECEIVER

B

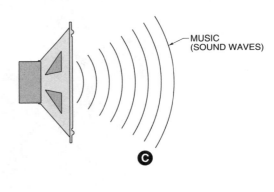

MUSIC (SOUND WAVES)

C

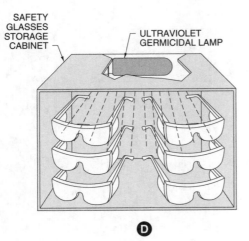

SAFETY GLASSES STORAGE CABINET

ULTRAVIOLET GERMICIDAL LAMP

D

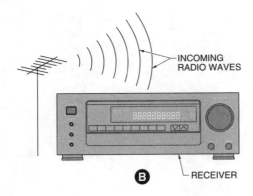

250 W HIGH-PRESSURE SODIUM LAMP

E

Basic Quantities

Activities

2

Name _____ Date _____

Interactive Activity Objectives

- Determine the specified power supply requirements for different models of an electrical device.
- Determine the specified output current rating for mechanical contacts and solid-state contacts for different models of an electrical device.

Procedure 2-1

Understanding basic electrical quantities such as voltage and current ratings is required when selecting and using electrical equipment. All electrical equipment has ratings that specify where a device can and cannot be used. Click on the TECO SG2 Client Simulation Software button on the CD-ROM. Install the software on your computer. Open a new ladder diagram document by clicking on the New Ladder Diagram icon, located directly below the File heading. Open a new circuit by clicking on the New Circuit icon, located directly below the File heading. The Select Model Type screen will appear. Answer the following questions using the technical data listed for the first four models.

_____ 1. What is the listed power-supply voltage level for Model SG2-10HR-A?

_____ 2. What is the listed power-supply voltage type for Model SG2-10HR-A?

_____ 3. What is the output current rating for each of its four relay contacts?

_____ 4. What is the listed power-supply voltage level for Model SG2-12HR-D?

_____ 5. What is the listed power-supply voltage type for Model SG2-12HR-D?

_____ 6. What is the output current rating for each of its four relay contacts?

_____ 7. What is the listed power-supply voltage level for Model SG2-12HR-12D?

_____ 8. What is the listed power-supply voltage type for Model SG2-12HR-12D?

_____ 9. What is the output current rating for each of its four relay contacts?

_____ 10. What is the listed power-supply voltage level for Model SG2-12HT-D?

_____ 11. What is the listed power-supply voltage type for Model SG2-12HT-D?

_____ 12. What is the output current rating for each of its four relay contacts?

Color Rendering Activity Objectives

• Observe how different lamp types render colors when illuminated indoors.

Procedure 2-2

Color rendering is the appearance of color when illuminated by a light source. For example, a red color may be rendered light, dark, pinkish, or yellowish depending on the light source under which it is viewed. Some lamp types have poor color rendering but have a high light output (lumens) per watt ratio. Other lamp types are manufactured to have different color rendering characteristics. With lighting, there is a tradeoff between good color rendering (true colors) and the amount of light delivered per watt. Click on the Color Rendering Simulation button on the CD-ROM. Click on each lamp type to view the application under the different lamp types.

LAMP CHARACTERISTICS SUMMARY				
Lamp	Lm/W	Rated Bulb Life*	Color Rendition	Operating Cost
Incandescent	15–25	750–1000	Excellent	Very high
Tungsten-halogen	20–25	1500–2000	Excellent	High
Fluorescent	55–100	7500–24,000	Very good	Average
Low-pressure sodium	190–200	1800	Poor	Low
Mercury-vapor	50–60	16,000–24,000	Depends on type used	Average
Metal-halide	80–125	3000–20,000	Very good	Average
High-pressure sodium	65–115	7500–14,000	Good (golden white)	Low

* in hours

1. Which lamp type produces the best color rendering for the given application?

2. Which lamp type produces a satisfactory color rendering for the given application and delivers the most light per watt (Lm/W)?

Ohm's Law and the Power Formula

Review Questions

3

Name _____ Date _____

True-False

T   F   **1.** Ohm's law is the relationship between voltage, current, and power in a circuit.

T   F   **2.** Circuit resistance typically decreased due to poor connections, loose connections, corrosion, or damaged components.

T   F   **3.** The power formula states that power in a circuit is equal to voltage times resistance.

T   F   **4.** Ohm's law states that resistance in a circuit is equal to voltage times current.

T   F   **5.** The power formula states that if the voltage in a circuit remains constant and the power required from the circuit changes, the current in the circuit remains constant.

T   F   **6.** Ohm's law states that if the resistance in a circuit remains constant, a change in current is directly proportional to a change in voltage.

T   F   **7.** The power required from a circuit changes any time loads are added (power increase) or removed (power decrease).

T   F   **8.** Ohm's law and the power formula are limited to circuits in which electrical resistance is the only significant opposition to the flow of current.

Multiple Choice

_____ **1.** Changing a circuit's resistance changes the amount of ___ flowing through the circuit.
    A. current
    B. voltage
    C. power
    D. energy

_____ **2.** The power formula states that voltage in a circuit is equal to power ___ current.
    A. divided by
    B. added to
    C. subtracted from
    D. multiplied by

21

_____ 3. Ohm's law and the power formula may be combined mathematically and written
as a combination of voltage and ___.
    A. current
    B. resistance
    C. power
    D. all of the above

_____ 4. Ohm's law states that voltage in a circuit is equal to resistance ___ current.
    A. times
    B. plus
    C. minus
    D. divided by

_____ 5. In circuits that contain ___ or capacitance, the opposition to the flow of current
is reactance.
    A. resistance
    B. inductance
    C. impedance
    D. voltage

Completion

_____ 1. Ohm's law states that ___ in a circuit is proportional to the voltage and inversely
proportional to the resistance.

_____ 2. Ohm's law states that current in a circuit is equal to voltage divided by ___.

_____ 3. The ___ is the relationship between power, voltage, and current in an electrical
circuit.

_____ 4. The power formula states that current in a circuit is equal to ___ divided by voltage.

_____ 5. Ohm's law states that if the ___ in a circuit remains constant, a change in resistance
produces an inversely proportional change in current.

Ohm's Law and Power Formula Calculations

_____ **1.** $E_T = \_\_$ V

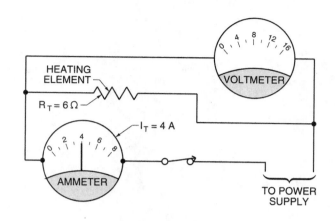

_____ **2.** $I_T = \_\_$ A

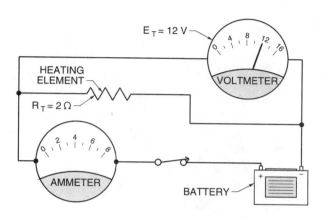

_____ **3.** $R_T = \_\_$ Ω

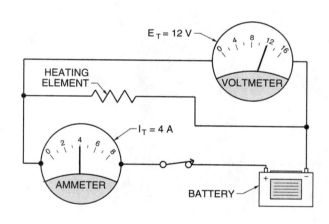

_____ **4.** $P_T = \_\_$ W

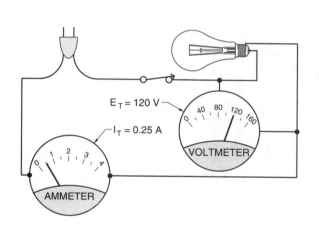

_____ **5.** $E_T = \_\_$ V

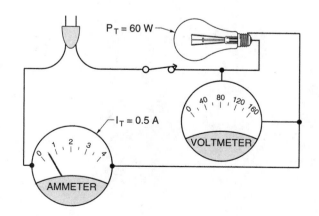

_____ **6.** $I_T = \_\_$ A

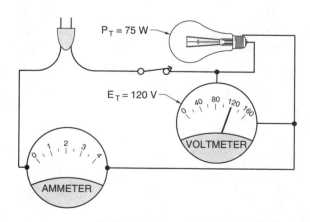

Name _____ **Date** _____

Determining Current, Resistance, Voltage, and Power

_____ **1.** $I_T =$ ___ mA

$E_T = 115\ V$ $R_T = 1.25\ k\Omega$

_____ **2.** $I_T =$ ___ μA

$E_T = 220\ V$ $R_T = 0.7\ M\Omega$

_____ **3.** $R_T =$ ___ Ω

$E_T = 24\ V$ $I_T = 80\ mA$

_____ **4.** $R_T =$ ___ Ω

$E_T = 36\ V$ $I_T = 1.4\ A$

_____ **5.** $E_T =$ ___ V

$I_T = 4.8\ mA$ $R_T = 25\ k\Omega$

_____ **6.** $E_T =$ ___ V

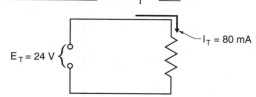

$I_T = 48\ mA$ $R_T = 750\ \Omega$

_____ **7.** $E_T =$ ___ V

$I_T = 240\ \mu A$ $R_T = 1\ M\Omega$

_____ **8.** $P_T =$ ___ mW

$E_T = 12\ V$ $R_T = 250\ \Omega$

_____ **9.** $P_T =$ ___ W

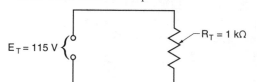

$E_T = 115\ V$ $R_T = 1\ k\Omega$

_____ **10.** $P_T =$ ___ mW

$E_T = 36\ V$ $R_T = 8\ k\Omega$

Name _____ Date _____

Determining Circuit Current

Complete the table with the required current values.

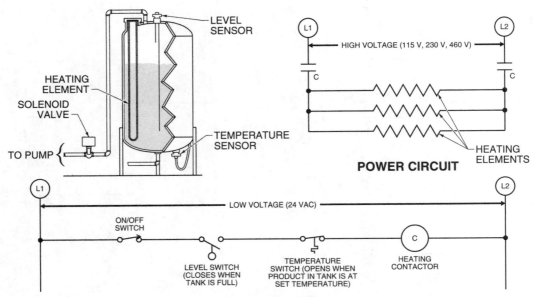

HEATING ELEMENT RATING			
Power Rating*	**Circuit Current****		
	115 V	**230 V**	**460 V**
1.5			
3			
5			
10	————		
12.5	————		
21.25	————		
27.8	————	————	
30	————	————	
50	————	————	

* in kW
** in A

Ohm's Law and the Power Formula

Worksheet 3-3

Name _____ Date _____

Applying Ohm's Law and the Power Formula

_____ **1.** $I_T =$ ___ A

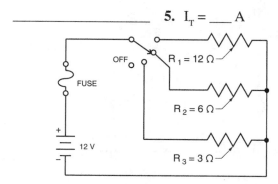

120 V SUPPLY
$R_T = 1.25 \text{ k}\Omega$

_____ **2.** $R_T =$ ___ Ω

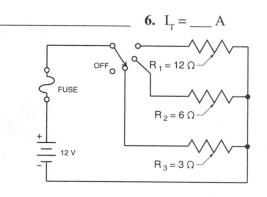

SOLID STRIP HEATER
31.5_A
240 V SUPPLY

_____ **3.** $E_T =$ ___ V

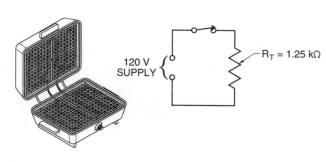

FINNED STRIP HEATER
8.32_A
$R_T = 25 \Omega$

_____ **4.** $P_T =$ ___ W

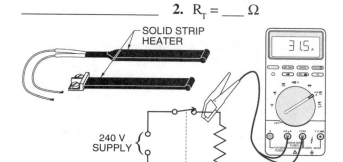

CONTROL SWITCH 1
115 V
230 V SUPPLY
115 V
$R_1 = 25 \Omega$
$R_2 = 25 \Omega$

_____ **5.** $I_T =$ ___ A

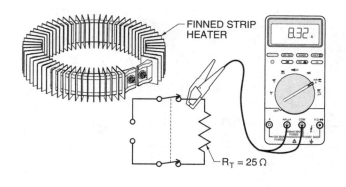

OFF
FUSE
$R_1 = 12 \Omega$
$R_2 = 6 \Omega$
12 V
$R_3 = 3 \Omega$

_____ **6.** $I_T =$ ___ A

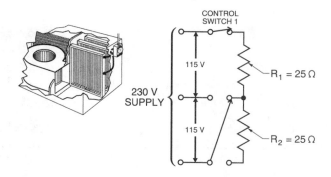

OFF
FUSE
$R_1 = 12 \Omega$
$R_2 = 6 \Omega$
12 V
$R_3 = 3 \Omega$

Name _____ Date _____

Determining Conductor Current

Determine the current in each conductor when all loads are ON.

_____ **1.** Current at A = ___ A

_____ **2.** Current at B = ___ A

_____ **3.** Current at C = ___ A

_____ **4.** Current at D = ___ A

_____ **5.** Current at E = ___ A

_____ **6.** Current at F = ___ A

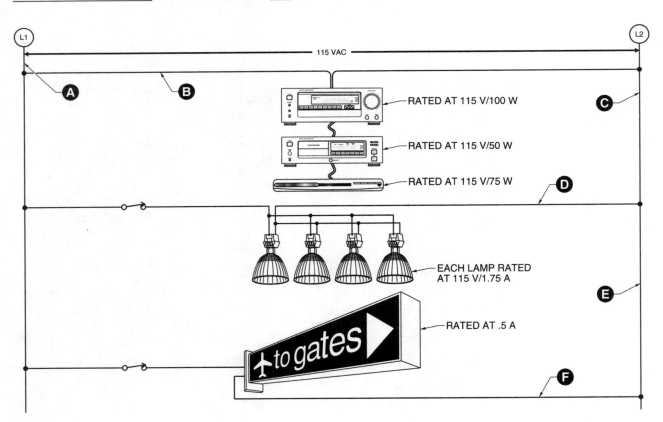

28

Ohm's Law and the Power Formula

Activities

3

Name _____ Date _____

Virtual Meter Activity Objectives

- Use a virtual DMM to show how voltage, current, and resistance measurements can be taken.

Procedure 3-1

Digital multimeters (DMMs) are used to take electrical measurements. Click on the Virtual Meters button on the CD-ROM. Next, click on the Virtual DMM 87V Demo button and answer the following questions.

1. Using the VDC and mVDC range, list the five VDC voltage ranges provided with this model DMM.

_____ 2. Using the ohms (Ω) range, what is the maximum resistance measurement this DMM can measure?

_____ 3. What is the meter setting that produces an audible signal when measuring continuity?

_____ 4. Using the highest current-range setting, what is the highest amount of current this DMM can measure?

5. What happens to the meter test leads when the meter is switched from the VDC (or Ω) setting to the current-range setting?

Calculation Activity Objectives

- Calculate the amount of circuit current by applying Ohm's law.
- Calculate the amount of circuit power by applying the power formula.

Procedure 3-2

Calculate the circuit current by applying Ohm's law.

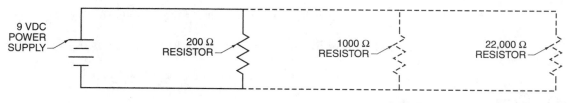

9 VDC POWER SUPPLY   200 Ω RESISTOR   1000 Ω RESISTOR   22,000 Ω RESISTOR

_____ **1.** The current in a circuit that contains a 9 VDC power supply and a 200 Ω resistor is ___ A.

_____ **2.** The current in a circuit that contains a 9 VDC power supply and a 1000 Ω resistor is ___ A.

_____ **3.** The current in a circuit that contains a 9 VDC power supply and a 22,000 Ω resistor is ___ A.

Calculate the circuit power by applying the power formula.

_____ **4.** The power in a circuit that contains a 9 VDC power supply and a 200 Ω resistor is ___ W.

_____ **5.** The power in a circuit that contains a 9 VDC power supply and a 1000 Ω resistor is ___ W.

_____ **6.** The power in a circuit that contains a 9 VDC power supply and a 22,000 Ω resistor is ___ W.

Measurement Activity Objectives

- Measure the amount of circuit current using a digital multimeter.
- Calculate the actual amount of circuit power using the power formula and the measured current values.

Procedure 3-3

Using the Electrical Component Identification card in the Components Kit, identify the 9 VDC battery, 200 Ω resistor, 1000 Ω resistor, and the 22,000 Ω resistor. Build the following circuit starting with the 1000 Ω resistor. Measure the current of each circuit by applying the In-Line Ammeter—DC Measurement procedure in the Appendix.

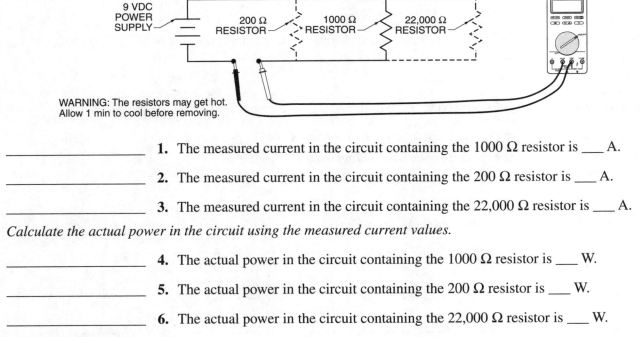

WARNING: The resistors may get hot.
Allow 1 min to cool before removing.

_____ **1.** The measured current in the circuit containing the 1000 Ω resistor is ___ A.

_____ **2.** The measured current in the circuit containing the 200 Ω resistor is ___ A.

_____ **3.** The measured current in the circuit containing the 22,000 Ω resistor is ___ A.

Calculate the actual power in the circuit using the measured current values.

_____ **4.** The actual power in the circuit containing the 1000 Ω resistor is ___ W.

_____ **5.** The actual power in the circuit containing the 200 Ω resistor is ___ W.

_____ **6.** The actual power in the circuit containing the 22,000 Ω resistor is ___ W.

Name _____ **Date** _____

True-False

T   F   **1.** The purpose of the NEC® is the practical safeguarding of persons and property from the hazards arising from the use of electricity.

T   F   **2.** NFPA is an independent organization that tests equipment and products to see if they conform to national codes and standards.

T   F   **3.** Improper electrical wiring or misuse of electricity causes destruction of equipment and fire damage to property.

T   F   **4.** The severity of an electrical shock depends on the amount of electric voltage that flows through the body.

T   F   **5.** More than 100,000 people are killed in electrical fires each year.

T   F   **6.** An enclosure is selected based on the location of the equipment and NEC® requirements.

T   F   **7.** Per OSHA standards, equipment is locked out and tagged out after any preventive maintenance or servicing is performed.

T   F   **8.** An electric arc is a discharge of electric current across an air gap.

T   F   **9.** Two areas requiring attention when working with electric motors are the electrical circuit and rotating shaft.

T   F   **10.** Heat stroke is much less serious than heat exhaustion.

T   F   **11.** Proper grounding of electrical tools, machines, equipment, and delivery systems is one of the most important factors in preventing hazardous conditions.

T   F   **12.** Class A fires include burning oil, gas, grease, paint, and other liquids that convert to a gas when heated.

Multiple Choice

_____ **1.** A(n) ___ is an accepted reference or practice.
   A. code
   B. standard
   C. agreement
   D. contract

_____ **2.** ___ is a national organization that helps identify industrial and public needs for standards.
   A. CSA
   B. OSHA
   C. NEMA
   D. ANSI

_____ **3.** ___ is a national organization that assists with information and standards concerning proper selection, ratings, construction, testing, and performance of electrical equipment.
   A. CSA
   B. OSHA
   C. NEMA
   D. ANSI

_____ **4.** Class ___ protective helmets protect against high-voltage shock and burns, impact hazards, and penetration by falling or flying objects.
   A. A
   B. B
   C. C
   D. X

_____ **5.** ___ is the most visible result of an injury.
   A. Shock
   B. Bleeding
   C. Heat stroke
   D. Poisoning

Completion

_____ **1.** A(n) ___ is a regulation or minimum requirement.

_____ **2.** Use only a Class ___ fire extinguisher on burning electrical equipment.

_____ **3.** ___ is the connection of all exposed noncurrentcarrying metal parts to the earth.

_____ **4.** ___ is the process of removing the source of electrical power and installing a lock which prevents the power from being turned ON.

_____ 5. ___ is the hazardous location category that refers to the generic hazardous material present.

_____ 6. ___ is the process of placing a danger tag on the source of electrical power that indicates that the equipment may not be operated until the danger tag is removed.

_____ 7. ___ is the hazardous location category that refers to the probability that a hazardous material is present.

_____ 8. An electrical ___ is a housing that protects wires and equipment and prevents personnel injury from accidental contact with a live circuit.

_____ 9. Lockout ___ are lightweight enclosures that allow the lockout of standard control devices.

_____ 10. ___ provide a written record of the steps taken to comply with industry safety standards as well as a method of tracking progress.

_____ 11. A(n) ___ is an eye and face protection device that covers the entire face with a plastic shield and is used for protection from flying objects.

_____ 12. A(n) ___ is an ear protection device worn over the ears.

_____ 13. The primary purpose of ___ and leather protectors is to insulate hands and lower arms from possible contact with live conductors.

_____ 14. ___ is help for a victim immediately after an injury and before professional medical help arrives.

_____ 15. The ___ approach boundary is the distance from an exposed energized conductor or circuit part where an increased risk of electric shock exists due to the close proximity of the person to the energized conductor or circuit part.

Name _____ Date _____

Fire Extinguisher Classes

_____ **1.** Class ___ extinguisher required for fire 1.

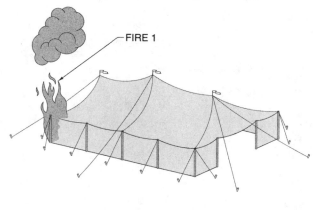

_____ **2.** Class ___ extinguisher required for fire 2.

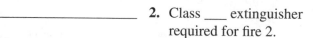
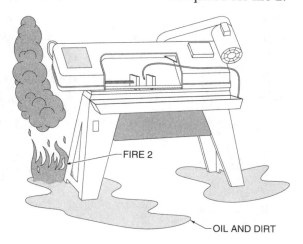

_____ **3.** Class ___ extinguisher required for fire 3.

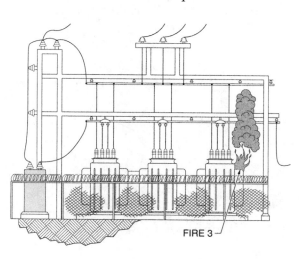

_____ **4.** Class ___ extinguisher required for fire 4.

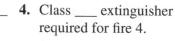
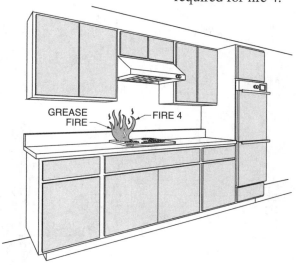

Name _____ **Date** _____

Lockout/Tagout

Determine if a lockout/tagout device is required or cannot be used for each service call.

_____ **1.** Lockout/tagout device ___.

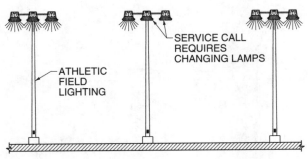

SERVICE CALL REQUIRES CHANGING LAMPS

ATHLETIC FIELD LIGHTING

_____ **2.** Lockout/tagout device ___.

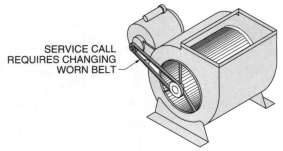

SERVICE CALL REQUIRES CHANGING WORN BELT

_____ **3.** Lockout/tagout device ___.

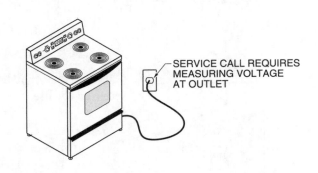

SERVICE CALL REQUIRES MEASURING VOLTAGE AT OUTLET

_____ **4.** Lockout/tagout device ___.

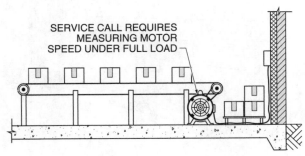

CANOPY

SERVICE CALL REQUIRES FAN REPLACEMENT

_____ **5.** Lockout/tagout device ___.

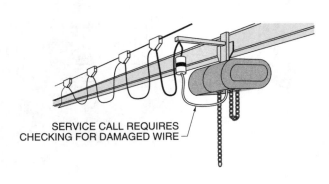

SERVICE CALL REQUIRES MEASURING MOTOR SPEED UNDER FULL LOAD

_____ **6.** Lockout/tagout device ___.

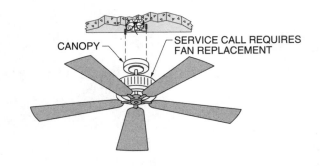

SERVICE CALL REQUIRES CHECKING FOR DAMAGED WIRE

Name _____ Date _____

Enclosures

_____ **1.** Required enclosure is NEMA type ___.

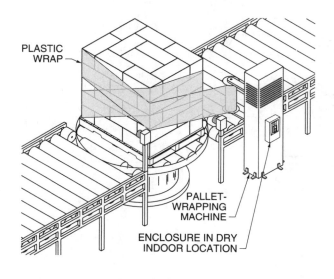

PLASTIC WRAP

PALLET-WRAPPING MACHINE

ENCLOSURE IN DRY INDOOR LOCATION

_____ **2.** Required enclosure is NEMA type ___.

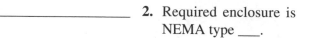
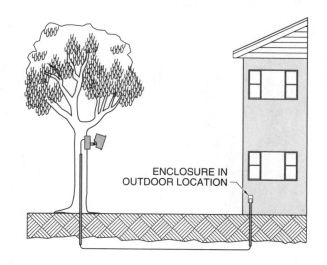

ENCLOSURE IN OUTDOOR LOCATION

_____ **3.** Required enclosure is NEMA type ___.

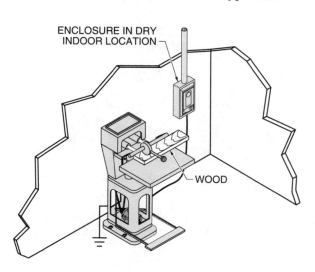

ENCLOSURE IN DRY INDOOR LOCATION

WOOD

_____ **4.** Required enclosure is NEMA type ___.

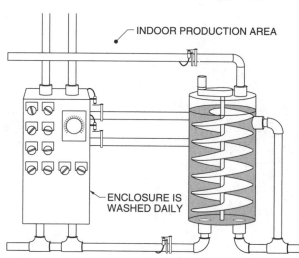

INDOOR PRODUCTION AREA

ENCLOSURE IS WASHED DAILY

Name _____ **Date** _____

Hazardous Locations

_____ **1.** Class ___

_____ **2.** Division ___

_____ **3.** Group ___

_____ **4.** Class ___

_____ **5.** Division ___

_____ **6.** Group ___

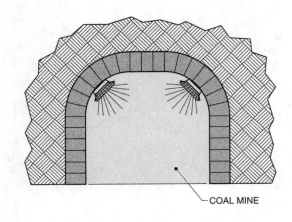

COAL MINE

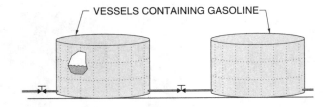

VESSELS CONTAINING GASOLINE

_____ **7.** Class ___

_____ **8.** Division ___

_____ **9.** Group ___

_____ **10.** Class ___

_____ **11.** Division ___

_____ **12.** Group ___

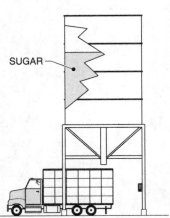

SUGAR

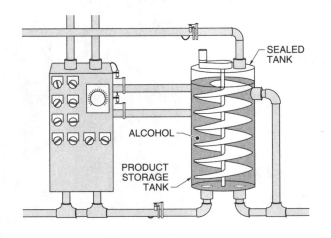

SEALED TANK

ALCOHOL

PRODUCT STORAGE TANK

Safety

Activities

4

Name _____ **Date** _____

Virtual Meter Activity Objectives

- Determine if electrical gloves are required during a given electrical measurement.
- Determine if the level of measured current can cause an electrical shock.

Procedure 4-1

Click on the Virtual Meters button on the CD-ROM. Next, click on the Virtual DMM 87V Demo button. Move the position of the selector switch from the OFF position to the VAC position (first position past OFF).

1. Based on the displayed meter measurement, should electrical gloves be worn when taking this measurement? Why or why not?

Move the position of the selector switch from the VAC position to the VDC position (next position).

2. Based on the displayed meter measurement, should electrical gloves be worn when taking this measurement? Why or why not?

Move the position of the selector switch from the VDC position to the mVDC position (next position).

3. Based on the displayed meter measurement, should electrical gloves be worn when taking this measurement? Why or why not?

Move the position of the selector switch from the mVDC position to the resistance (Ω) position (next position).

4. Based on the displayed meter measurement, should electrical gloves be worn when taking this measurement? Why or why not?

Move the position of the selector switch from the resistance position to the diode test position (next position).

5. Based on the displayed meter measurement, should electrical gloves be worn when taking this measurement? Why or why not?

Move the position of the selector switch from the diode test position to the A/mA AC (current-measuring) position (next position).

6. Based on the displayed meter measurement, is the current level high enough to cause an electrical shock?

Move the position of the selector switch from the A/mA AC test position to the μA AC (current-measuring) position (next position).

7. Based on the displayed meter measurement, is the current level high enough to cause an electrical shock?

Grounding Activity Objectives

• Review an example of the problems that a poorly grounded system caused.
• Understand the solution to problems that result from poor grounding.

Procedure 4-2

Click on the Grounding Application Videos button on the CD-ROM. Next, click on the Grounding Two Separate Systems button.

1. What are the two different types of grounding discussed?

Click on the Grounding Application Videos button on the CD-ROM. Next, click on the Improper Grounding Case Study buttons.

2. What problem was reported by the 911 center?

3. How much money was the 911 center spending to repair the problem?

4. Before the problem was corrected, what was the high-resistance value of some of the grounds?

5. After the problem was corrected, what was the resistance of the ground system?

6. How deep were the new ground rods driven?

Math Principles

Review Questions

5

Name _____ Date _____

True-False

T F **1.** Whole numbers (integers) are numbers that have no fractional or decimal parts.

T F **2.** The lowest common denominator is the smallest number into which the denominators of a group of two or more fractions divides an exact number of times.

T F **3.** A denominator is a number in a fraction that indicates how many of the whole are used.

T F **4.** An exponent is a number that indicates the number of times a base is raised to a power (number of times used as a factor).

T F **5.** A square root of a number is a number which, when multiplied by itself, gives the original number.

T F **6.** A ratio is the relationship between two quantities or terms.

T F **7.** A proportion is a number with a fixed value.

T F **8.** Diameter is the boundary of a circle.

Multiple Choice

_____ **1.** A(n) ___ fraction is a fraction with a denominator of 10, 100, 1000, etc.
  A. proper
  B. decimal
  C. improper
  D. mixed

_____ **2.** A(n) ___ number is any number that cannot be divided by 2 an exact number of times.
  A. even
  B. transverse
  C. odd
  D. chromatic

41

_____ 3. Fractions must have common ___ before they can be subtracted.
  A. numerators
  B. decimals
  C. denominators
  D. powers

_____ 4. ___ are two or more numbers multiplied together to give a product.
  A. Exponents
  B. Fractions
  C. Factors
  D. Roots

_____ 5. A(n) ___ proportion is a statement of equality between two ratios in which the first of four terms divided by the second equals the third divided by the fourth.
  A. direct
  B. inverse
  C. compound
  D. variable

Completion

_____ 1. ___ is the science of numbers and their operations, interrelations, and combinations.

_____ 2. A(n) ___ is any number associated with a point on a line.

_____ 3. A(n) ___ is a part of a whole unit or number.

_____ 4. ___ is the branch of mathematics that involves the computation (addition, subtraction, multiplication, and division) of positive real numbers.

_____ 5. The ___ of a fraction is a number that indicates the number of equal parts into which a unit is divided.

_____ 6. The value of a fraction is unchanged when a fraction is multiplied or divided by ___.

_____ 7. A(n) ___ fraction is a fraction that has a numerator smaller than its denominator.

_____ 8. A(n) ___ decimal number is a decimal number that has a whole number and a decimal number separated by a decimal point.

_____ 9. A(n) ___ fraction is a fraction that has a numerator larger than its denominator.

_____ 10. ___ is the process of obtaining a number to a required degree of accuracy.

_____ 11. A(n) ___ is a fraction that has a denominator of 100.

_____ 12. ___ is the number of unit squares equal to the surface of an object.

_____ 13. A negative base number raised to an odd power becomes a(n) ___ integer.

_____ **14.** A negative base number raised to an even power becomes a(n) ___ integer.

_____ **15.** A(n) ___ is a diagram that shows the continuous relationship between two or more variables.

_____ **16.** A(n) ___ is an expression indicating that two ratios are equal.

_____ **17.** A(n) ___ proportion is a statement of equality between two ratios in which an increase in one quantity results in a proportional decrease in the other related quantity.

_____ **18.** A(n) ___ proportion is a proportion in which some terms are products of two variables.

_____ **19.** A(n) ___ is a factor that is multiplied by the base number.

_____ **20.** The two most commonly used systems of measurement are the English system and the ___ system.

Math Symbols

_____ **1.** Square root

_____ **2.** Infinity

_____ **3.** Perpendicular

_____ **4.** Right angle

_____ **5.** Therefore

_____ **6.** Is approximately equal to

_____ **7.** Parallel

_____ **8.** Not equal to

_____ **9.** Angle

_____ **10.** Pi

$$\neq$$
Ⓐ

$$\approx$$
Ⓑ

$$\|$$
Ⓒ

$$\therefore$$
Ⓓ

$$\angle$$
Ⓔ

$$\pi$$
Ⓕ

$$\perp$$
Ⓖ

$$\llcorner$$
Ⓗ

$$\sqrt{}$$
Ⓘ

$$\infty$$
Ⓙ

Name _____ Date _____

Conversions

_____ **1.** The floor area of grain elevator A is ___ sq ft.

_____ **2.** The circumference of grain elevator A is ___′.

_____ **3.** Grain elevator A holds ___ cu yd of grain.

_____ **4.** An electric grain dryer that can process (dry) 3000 cu ft per hour must run ___ hr to process one full elevator of grain. (Round to nearest full hour.)

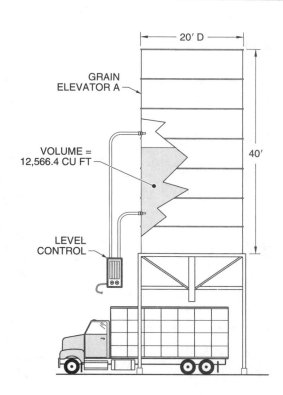

GRAIN ELEVATOR A

VOLUME = 12,566.4 CU FT

LEVEL CONTROL

20′ D

40′

_____ **5.** Tank B holds ___ gal. of oil.

_____ **6.** A ½ HP motor/pump rated to pump 50 gpm to a height of 25′ must run ___ hr to fill the tank.

_____ **7.** A(n) ___ HP motor/pump is required to fill the tank in approximately 1 hr. *Note:* Assume motors/pumps are available in ½ HP increments and round the motor/pump size to the next ½ HP larger size.

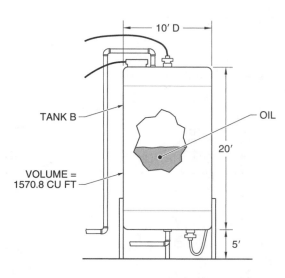

TANK B

OIL

VOLUME = 1570.8 CU FT

10′ D

20′

5′

Math Principles

Worksheet 5-2

Name _____ Date _____

Graphs

_____ **1.** A solid-state device in a 70°C location can control ____ A of load current when no heat sink is used.

_____ **2.** A solid-state device in a 70°C location can control ____ A of load current when heat sink 3 is used.

_____ **3.** A solid-state device in a 50°C location can control ____ A of load current when heat sink 2 is used.

_____ **4.** A solid-state device in a 100°C location can control ____ A of load current when heat sink 2 is used.

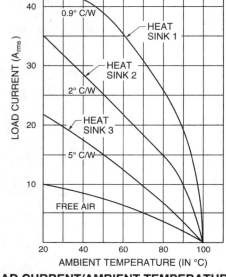

LOAD CURRENT/AMBIENT TEMPERATURES

_____ **5.** A 10 A rated heater trips at ____ A in an ambient temperature of 45°C.

_____ **6.** A 10 A rated heater trips at ____ A in an ambient temperature of 35°C.

_____ **7.** A 10 A rated heater trips at ____ A in an ambient temperature of 60°C.

_____ **8.** A 10 A rated heater trips at ____ A in an ambient temperature of 20°C.

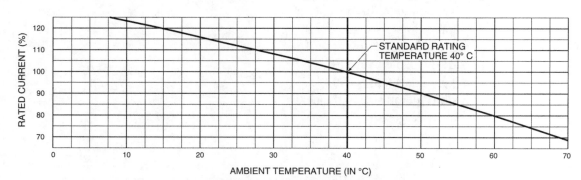

HEATER AMBIENT TEMPERATURE CORRECTION

Math Principles

Worksheet 5-3

Name _____ Date _____

Fractions

_____ **1.** Total horsepower load at the main disconnect switch = ___ HP.

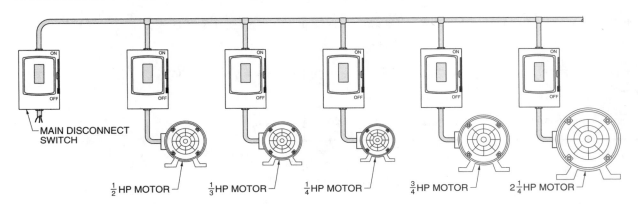

MAIN DISCONNECT SWITCH

$\frac{1}{2}$ HP MOTOR $\frac{1}{3}$ HP MOTOR $\frac{1}{4}$ HP MOTOR $\frac{3}{4}$ HP MOTOR $2\frac{1}{4}$ HP MOTOR

_____ **2.** Tension tester should measure ___ /64 for proper belt tension.

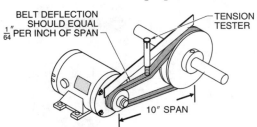

BELT DEFLECTION SHOULD EQUAL $\frac{1}{64}''$ PER INCH OF SPAN — TENSION TESTER

10" SPAN

_____ **3.** Tension tester should measure ___ /64 for proper belt tension.

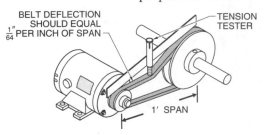

BELT DEFLECTION SHOULD EQUAL $\frac{1}{64}''$ PER INCH OF SPAN — TENSION TESTER

1' SPAN

_____ **4.** Tension tester should measure ___ /64 for proper belt tension.

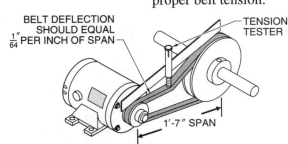

BELT DEFLECTION SHOULD EQUAL $\frac{1}{64}''$ PER INCH OF SPAN — TENSION TESTER

1'-7" SPAN

_____ **5.** Tension tester should measure ___ /64 for proper belt tension.

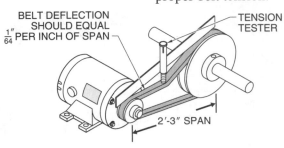

BELT DEFLECTION SHOULD EQUAL $\frac{1}{64}''$ PER INCH OF SPAN — TENSION TESTER

2'-3" SPAN

Name _____ **Date** _____

Percentages

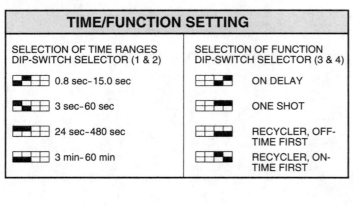

TIME/FUNCTION SETTING	
SELECTION OF TIME RANGES DIP-SWITCH SELECTOR (1 & 2)	SELECTION OF FUNCTION DIP-SWITCH SELECTOR (3 & 4)
0.8 sec–15.0 sec	ON DELAY
3 sec–60 sec	ONE SHOT
24 sec–480 sec	RECYCLER, OFF-TIME FIRST
3 min–60 min	RECYCLER, ON-TIME FIRST

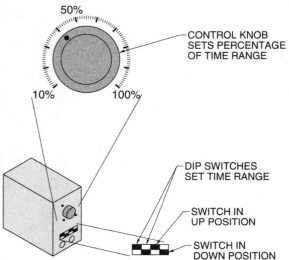

50%

CONTROL KNOB SETS PERCENTAGE OF TIME RANGE

10%   100%

DIP SWITCHES SET TIME RANGE

SWITCH IN UP POSITION

SWITCH IN DOWN POSITION

_____ **1.** Timer setting = ___ sec

70%

_____ **2.** Timer setting = ___ sec

60%

_____ **3.** Timer setting = ___ sec

20%

_____ **4.** Timer setting = ___ sec

90%

_____ **5.** Timer setting = ___ sec

40%

_____ **6.** Timer setting = ___ sec

30%

_____ **7.** Timer setting = ___ sec

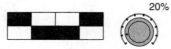

15%

_____ **8.** Timer setting = ___ sec

45%

Math Principles

Worksheet 5-5

Name _____ **Date** _____

Prefixes

_____ **1.** 12,000 Ω = ___ kΩ

_____ **2.** 9,872,000 W = ___ MW

_____ **3.** 0.00001 A = ___ μA

_____ **4.** 25,500 Ω = ___ kΩ

_____ **5.** 0.00005 A = ___ μA

_____ **6.** 0.0000005 = ___ μA

_____ **7.** 1,500,000 Ω = ___ MΩ

_____ **8.** 0.010005 A = ___ mA

_____ **9.** 62,000 W = ___ kW

_____ **10.** 112.3 V = ___ mV

_____ **11.** 0.0505 A = ___ mA

_____ **12.** 1,000,100 W = ___ kW

_____ **13.** 0.1234 A = ___ mA

_____ **14.** 0.0101 A = ___ mA

_____ **15.** 1.705 V = ___ mV

_____ **16.** 0.00000013 A = ___ mA

_____ **17.** 10 A = ___ kA

_____ **18.** 170,000 Ω = ___ kΩ

_____ **19.** 23.5 V = ___ mV

_____ **20.** 1 W = ___ mW

_____ **21.** 12 mA = ___ A

_____ **22.** 30.3 μA = ___ A

_____ **23.** 1.5 kΩ = ___ Ω

_____ **24.** 105 mA = ___ A

_____ **25.** 1.2 MV = ___ V

_____ **26.** 15.5 kW = ___ W

_____ **27.** 0.02 kW = ___ W

_____ **28.** 1.3 MΩ = ___ Ω

_____ **29.** 10.3 μA = ___ A

_____ **30.** 0.5 mΩ = ___ Ω

_____ **31.** 67 kΩ = ___ Ω

_____ **32.** 0.05 mV = ___ V

_____ **33.** 67.5 kΩ = ___ Ω

_____ **34.** 0.055 mA = ___ A

_____ **35.** 72 mV = ___ V

_____ **36.** 1.3 kV = ___ V

_____ **37.** 101.1 kW = ___ W

_____ **38.** 1.72 MΩ = ___ Ω

_____ **39.** 30 μA = ___ A

_____ **40.** 0.4 MΩ = ___ Ω

Name _____ Date _____

Metric System Conversions

_____ 1. Sensing distance = ___″

_____ 2. Maximum output = ___A

_____ 3. Minimum output = ___A

_____ 4. Leakage current = ___A

_____ 5. Type D proximity switch X dimension = ___″

_____ 6. Type D proximity switch Y dimension = ___″

SPECIFICATIONS	
SUPPLY VOLTAGE	10 VDC-40 VDC
AMBIENT TEMPERATURE	–20° C TO +60° C
TARGET	FERROUS METALS
SENSING DISTANCE	40 mm
LEAKAGE CURRENT	1.7 mA
RESPONSE TIME	150 Hz
OUTPUT	MINIMUM: 20 mA MAXIMUM: 200 mA

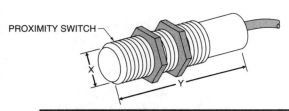

PROXIMITY SWITCH

PROXIMITY SWITCH SIZES*							
DIMENSIONS	TYPE						
	A	B	C	D	E	F	G
X	6	8	12	16	18	20	30
Y	35	72	42	80	42	80	50

* in mm

_____ 7. Belt width of ⅜″ = ___ cm

_____ 8. Belt width of ⅜″ = ___ mm

_____ 9. Belt width of ½″ = ___ cm

_____ 10. Belt width of ½″ = ___ mm

_____ 11. Belt width of ²¹⁄₃₂″ = ___ cm

_____ 12. Belt width of ²¹⁄₃₂″ = ___ mm

_____ 13. 0.76 kW = ___ W

_____ 14. 2.5 kW = ___ W

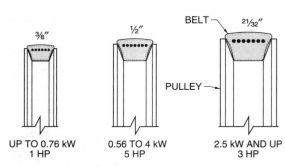

⅜″	½″	BELT ²¹⁄₃₂″
UP TO 0.76 kW 1 HP	0.56 TO 4 kW 5 HP	PULLEY 2.5 kW AND UP 3 HP

V-BELT/MOTOR SIZE

Math Principles

Activities

5

Name _____ **Date** _____

Virtual Meter Activity Objectives

- Take small displayed meter readings that include the prefix milli and micro and convert each measurement to a base number.
- Take large displayed meter readings that include the prefix kilo and convert each measurement to a base number.

Procedure 5-1

Click on the Virtual Meters button on the CD-ROM. Next, click on the Virtual DMM 87V Demo button. Move the position of the selector switch to the mVDC position.

_____ **1.** The meter displays a measurement in mVDC. This measurement converted to a base number is ___ VDC.

Move the position of the selector switch to the µA position. The orange button toggles the meter between AC and DC measurements.

_____ **2.** The displayed DC current measurement converted to a base number is ___ A.

_____ **3.** The displayed AC current measurement converted to a base number is ___ A.

Click on the Virtual Meters button on the CD-ROM. Next, click on the Virtual DMM 43B Demo button. Follow the directions provided to turn ON the meter. From the main menu, select the volts/amps/hertz screen and press enter.

_____ **4.** The meter displays the measured AC voltage in kVA. This measurement converted to a base number is ___ VAC.

From the main menu, select the power screen and press enter.

_____ **5.** The meter displays the measured power in kW and kVA. The highest displayed measurement converted to a base number is ___ W and ___ VA.

From the main menu, select the ohms (Ω) screen and press enter.

_____ **6.** The meter displays the measured resistance in kΩ. This measurement converted to a base number is ___ Ω.

Programming Activity Objectives

- Identify the math symbols used in simulation software when programming electrical circuits that use analog inputs and math comparative logic blocks.

Procedure 5-2

1. Open the TECO SG2 Client simulation software on the CD-ROM.

2. Open a new function block diagram document by clicking on the New FBD icon located directly under the Help icon.

3. Open a new circuit by clicking of the New Circuit icon located directly under the File icon at the top of the screen.

4. In the Select Type window, select an SG2-10HR-A model.

5. Open the function block (FB) folder by clicking on the FB icon at the bottom of the screen.

6. Open the analog (Gx) folder by clicking on the Gx icon at the bottom of the screen.

7. Place an analog function block in the middle of the screen. In the analog function block window, set the function mode to 2 and click ok.

8. Open the help folder by clicking on the question mark (?) icon at the top of the screen. Click on block instruction, function block, and then analog to obtain information on a mode 2 setting.

_____ 9. In mode 2, the requirement for the output of the function block to be ON occurs when input Ax is ___ or ___ to Ay.

_____ 10. In mode 3, the requirement for the output of the function block to be ON occurs when input Ax is ___ or ___ to Ay.

11. On the left side of the screen, place one standard ON/OFF input (I-01, labeled "enable") and three analog inputs (Ai-05 labeled "input Ax"; Ai-06 labeled "input Ay"; and Ai-07 labeled "reference") using the tool palette at the bottom of the screen. Inputs are found by opening the Co icon and clicking on I-IN for inputs. Analog inputs are found by opening the Co icon and clicking on A-IN for analog inputs.

12. Place one output (Q-01) on the right side of the screen using the tool palette at the bottom of the screen. Outputs are found by opening the Co icon and clicking on Q-OUT for outputs.

13. Connect inputs I-01, Ai-05, Ai-06, and Ai-07 to the four inputs of the analog function block in the order from top to bottom (I-01 on top and Ai-07 on bottom) and the output to the right side of the function block using the Connect icon at the bottom of the screen. Connect the lines as required.

14. Select the Simulator icon at the top of the screen (located to the left of the Run icon).

15. Test the circuit by using the Input Status Tool window and the Expanded Analog Input window. Set different values in analog inputs Ai-05 and Ai-06 (Ai-07 is not used in Mode 2). Enable the circuit by clicking on Input I-01.

16. End the simulation by clicking the Stop icon at the top of the screen.

17. Set the analog function block to Mode 3 and reset the circuit using the simulator.

18. End the simulation by clicking the Stop icon at the top of the screen.

19. Set the analog function block to Mode 1 and reset the circuit using the simulator. In this mode, the reference input (Ai-07) is used.

20. Print a copy of the circuit by clicking the Print icon at the top of the screen.

21. Save the program as required.

Math Applications

Review Questions

6

Name _____ Date _____

True-False

T	F	**1.** The amount of power required to heat air must be determined when selecting the correct heater size.
T	F	**2.** Air produces resistance against all moving objects.
T	F	**3.** One of the advantages of a variable frequency drive is it can be applied to standard 3ϕ induction motors.
T	F	**4.** Conductors are sized large enough to prevent no more than a 3% voltage drop at the farthest point per NEC® 210.19(A) Informational Note No. 4.
T	F	**5.** Proximity and photoelectric switches are used to detect moving objects without touching them.
T	F	**6.** A pulley can be used to change the output speed of a motor.
T	F	**7.** Air flow rate varies indirectly with the width of a blower impeller.
T	F	**8.** The cost of electricity is based on the number of kWh of electricity consumed and the utility rate.
T	F	**9.** The heating elements in an electric forced-air heating unit produce the same heat output if connected to a voltage supply greater or less than their rated voltage.
T	F	**10.** All conductors (wires) have resistance.
T	F	**11.** To select the correct size heater to heat water, the temperature rise that the heating element must produce and the amount of water to be heated must be determined.
T	F	**12.** AC induction motors operate at a constant speed for a given frequency and number of poles.
T	F	**13.** The longer a conductor run, the lower the total resistance.
T	F	**14.** Resistance can be reduced by using a good conductor and/or increasing the size (cross-sectional area) of the conductor.

53

T   F   **15.** A VFD can control the speed of a standard induction motor.

T   F   **16.** In most power measurements, the kilowatt-hour (kWh) is used instead of the watt-hour.

T   F   **17.** Electric heating elements are rated in wattage for a given applied voltage.

T   F   **18.** The VFD must be programmed to apply a boost voltage at high motor speeds to compensate for power losses at high speeds.

T   F   **19.** For any given conductor, the larger the wire's diameter, the greater the resistance.

T   F   **20.** An application's horsepower requirements must be determined when selecting a motor or motor drive (VFD) because motors and drives are horsepower rated.

T   F   **21.** When two speakers are connected in parallel, their total resistance increases.

T   F   **22.** Surge protectors are rated for the maximum current for which they are designed.

Multiple Choice

_____ **1.** Portable generators are rated by their ___.
     A. maximum power
     B. surge power
     C. voltage
     D. all of the above

_____ **2.** A ___ connection is a connection that has two or more components connected so that there is more than one path for current flow.
     A. parallel
     B. series
     C. vector
     D. all of the above

_____ **3.** ___ is the process of saving energy and reducing costs by minimizing the amount of power used.
     A. Impedance
     B. Power optimization
     C. Energy consumption
     D. Energy efficiency

_____ **4.** The major factor that affects air flow rate is fan ___.
     A. operating speed
     B. impeller diameter
     C. impeller width
     D. all of the above

_____ **5.** Impedance is measured in ___.
      A. kW
      B. Hz
      C. ohms
      D. rpm

Completion

_____ **1.** The higher the required temperature rise and/or the faster the airflow, the ___ the required heater size for a given application.

_____ **2.** The cross-sectional area of round conductors is measured in ___.

_____ **3.** For proper operation and future speed changes, a photoelectric or proximity switch with an operating frequency of at least ___% greater should be used for a given application.

_____ **4.** ___ is reduced by reducing the frontal area of the moving object and/or reducing the speed of the moving object.

_____ **5.** The V/Hz ratio is the ratio between the voltage applied to the stator and the ___ of the voltage applied to the stator of the motor.

_____ **6.** Strobe tachometers normally display the speed of moving objects in ___.

_____ **7.** The amount of energy used by an electric device and the amount charged for that use by a power company is expressed in ___.

_____ **8.** In most residential, commercial, and industrial locations, motors account for more than ___% of all power used.

_____ **9.** Proximity and photoelectric switches have a rated activating frequency which is often stated in ___.

_____ **10.** A(n) ___ is an electrical device that continuously monitors incoming power and automatically clamps any excess voltage that could cause damage.

_____ **11.** ___ is the total opposition of any combination of resistance, inductive reactance, or capacitive reactance offered to the flow of alternating current.

_____ **12.** A(n) ___ connection has two or more components connected so there is only one path for current flow.

Name _____ **Date** _____

Heating Water and Oil

_____ 1. ___ kW of electric power are required by a heater to raise the temperature of 500 gal. of water in a tank 80°F in 2 hr.

_____ 2. ___ kW of electric power are required by a heater to raise the temperature of 500 gal. of water in a tank 80°F in 5 hr.

_____ 3. ___ kW of electric power are required by a heater to raise the temperature of 500 gal. of water in a tank 25°F in 5 hr.

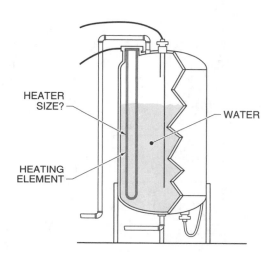

_____ 4. ___ kW of electric power are required by a heater to raise the temperature of 2000 gal. of oil in a tank 50°F in 8 hr.

_____ 5. ___ kW of electric power are required by a heater to raise the temperature of 20 gal. of oil in a tank 100°F in ½ hr.

_____ 6. ___ kW of electric power are required by a heater to raise the temperature of 20 gal. of oil in a tank 100°F in 4 hr.

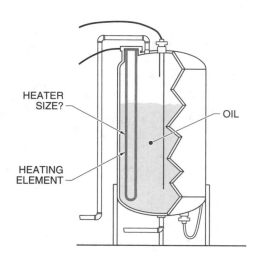

Name _____ Date _____

Heating Air and Electric Heater Output

_____ 1. ___ kW of electric power are required by a heater to raise the temperature of air 50°F with an airflow rate of 200 cfm.

_____ 2. ___ kW of electric power are required by a heater to raise the temperature of air 50°F with an airflow rate of 400 cfm.

_____ 3. ___ kW of electric power are required by a heater to raise the temperature of air 100°F with an airflow rate of 200 cfm.

_____ 4. The actual heater power output of a 5 kW, 480 V heating element connected to a 460 V supply is ___ W.

_____ 5. The actual heater power output of a 5 kW, 480 V heating element connected to a 230 V supply is ___ W.

_____ 6. The actual heater power output of a 5 kW, 480 V heating element connected to a 120 V supply is ___ W.

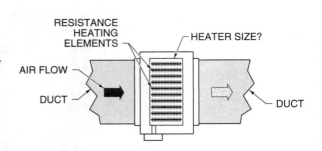

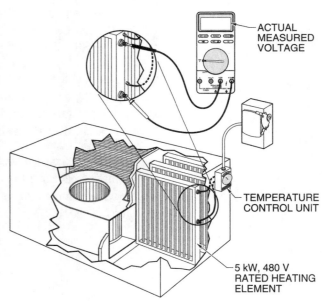

Name _____ Date _____

Conductor Maximum Length Calculation

_____ **1.** The maximum distance the load may be located from the power panel is ___'.

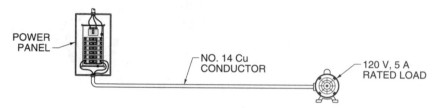

POWER PANEL — NO. 14 Cu CONDUCTOR — 120 V, 5 A RATED LOAD

_____ **2.** The maximum distance the load may be located from the power panel is ___'.

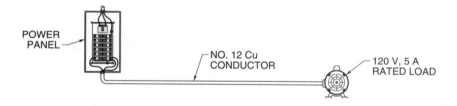

POWER PANEL — NO. 12 Cu CONDUCTOR — 120 V, 5 A RATED LOAD

_____ **3.** The maximum distance the load may be located from the power panel is ___'.

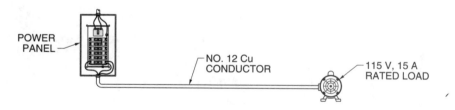

POWER PANEL — NO. 12 Cu CONDUCTOR — 115 V, 15 A RATED LOAD

_____ **4.** The maximum distance the load may be located from the power panel is ___'.

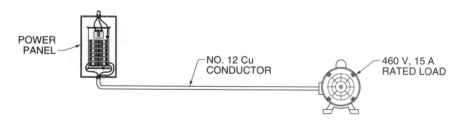

POWER PANEL — NO. 12 Cu CONDUCTOR — 460 V, 15 A RATED LOAD

Math Applications

Worksheet 6-4

Name _____ **Date** _____

Motor Required Horsepower and Pulley Size

_____ **1.** A(n) ___ HP motor is required to lift a 500 lb load vertically at a rate of 25 fpm.

_____ **2.** A(n) ___ HP motor is required to lift a 500 lb load vertically at a rate of 50 fpm.

_____ **3.** A(n) ___ HP motor is required to lift a 1000 lb load vertically at a rate of 50 fpm.

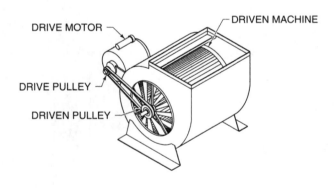

_____ **4.** A(n) ___″ driven pulley diameter is required if a motor running at 1800 rpm has a 2″ pulley and the driven machine is to be operated at 900 rpm.

_____ **5.** A(n) ___″ driven pulley diameter is required if a motor running at 1800 rpm has a 4″ pulley and the driven machine is to be operated at 900 rpm.

_____ **6.** A(n) ___″ driven pulley diameter is required if a motor running at 900 rpm has a 4″ pulley and the driven machine is to be operated at 1800 rpm.

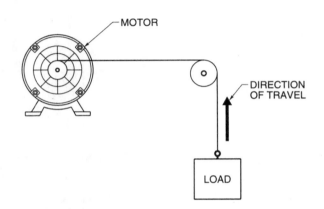

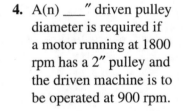

Name _____ Date _____

Object Activating Speed and Air Resistance

_____ 1. ___ ops occur when a strobe tachometer measures moving bottles at a rate of 50 fpm when the bottles are spaced 5 per foot.

_____ 2. ___ ops occur when a strobe tachometer measures moving bottles at a rate of 50 fpm when the bottles are spaced 4 per foot.

_____ 3. ___ ops occur when a strobe tachometer measures moving bottles at a rate of 100 fpm when the bottles are spaced 2 per foot.

_____ 4. The air resistance of a truck traveling at 25 mph and having a frontal area of 80 sq ft = ___ lb.

_____ 5. The air resistance of a truck traveling at 50 mph and having a frontal area of 80 sq ft = ___ lb.

_____ 6. The air resistance of a truck traveling at 50 mph and having a frontal area of 40 sq ft = ___ lb.

PROXIMITY SWITCH

BOTTLES

DIGITAL STROBOSCOPE
120
FPM

STROBE TACHOMETER

AIR RESISTANCE

MOVING TRUCK

Name _____ Date _____

Blower Sizing

_____ **1.** A blower delivering 400 cfm of air when operating at 1200 rpm delivers ___ cfm of air when its operating speed is increased to 1800 rpm.

_____ **2.** A blower delivering 400 cfm of air when operating at 1200 rpm delivers ___ cfm of air when its operating speed is increased to 900 rpm.

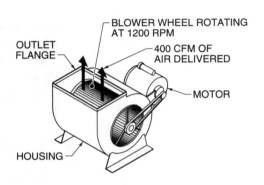

_____ **3.** A blower with a 3″ diameter blower wheel delivering 300 cfm of air delivers ___ cfm of air with a 4″ diameter blower wheel.

_____ **4.** A blower with a 3″ diameter blower wheel delivering 300 cfm of air delivers ___ cfm of air with a 8″ diameter blower wheel.

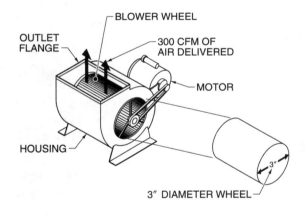

_____ **5.** A 10″ wide blower wheel delivers ___ cfm of air if a 5″ wide blower wheel delivers 300 cfm.

_____ **6.** A 20″ wide blower wheel delivers ___ cfm of air if a 5″ wide blower wheel delivers 300 cfm.

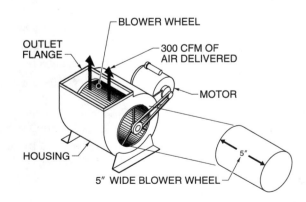

Math Applications

Activities

6

Name _____ **Date** _____

Virtual Meter Activity Objectives

- Understand how a digital meter that can measure rms AC voltage and peak AC voltage is used to capture and record a voltage surge.
- Use the information captured on a digital multimeter when troubleshooting a voltage surge problem.

Procedure 6-1

Click on the Virtual Meters button on the CD-ROM. Next, click on the Virtual DMM 87V Demo button. Move the selector switch to the VAC position and place the meter in the MIN/MAX recording mode.

_____ **1.** What is the listed minimum voltage?

_____ **2.** What is the listed maximum voltage?

_____ **3.** What is the listed average voltage?

_____ **4.** Based on the meter voltage reading (base reading, not MIN/MAX/AVE), what should the peak voltage be?

Place the meter in the peak measuring mode.

_____ **5.** What is the measured peak voltage?

If the peak voltage changes, the meter records a new peak value.

_____ **6.** What is the rated peak voltage capture time of the meter?

_____ **7.** This value converted to milliseconds is ___.

_____ **8.** This value converted to seconds is ___.

Graph Usage Activity Objectives

- Use a manufacturer specification chart to determine the amount of light reduction of a bulb over time.
- Use a manufacturer specification chart to determine the best operating position of a lamp bulb.

Procedure 6-2

Graphs are used by manufacturers to provide technical data that can be used when selecting, installing, and troubleshooting their products. An understanding of manufacturer charts is required when reading technical data bulletins. Click on the Lamp Data Sheets button on the CD-ROM. Next, click on the Compact Fluorescent Lamp button. Lamp bulbs produce their peak light output (amount of lumens) when first installed and have a reduced light output as the bulb ages.

_____ **1.** What is the approximate initial lumen output for a 26 W plug-in compact fluorescent lamp bulb?

_____ **2.** What is the approximate percentage reduction in lamp output after 6000 operating hours from the lamp output when the lamp was first installed (initial lumens)?

For each of the following, determine whether the bulb should be mounted in the up, down, or horizontal position.

_____ **3.** At approximately 86°F, a bulb should be mounted in the ___ position to produce close to the rated light output of the bulb.

_____ **4.** At approximately 5°C, a bulb should be mounted in the ___ position to produce close to the rated light output of the bulb.

_____ **5.** At approximately 20°C, a bulb should be mounted in the ___ position to produce close to the rated light output of the bulb.

Lamp Shipping Data Activity Objectives

- Use math principles to determine the weight of a pallet of lamp bulbs.
- Use math principles to determine the volume of a pallet of lamp bulbs.

Procedure 6-3

Click on the Lamp Data Sheets button on the CD-ROM. Next, click on the Energy Saver Dimmable Lamp button. A pallet of energy-saving dimmable fluorescent lamps is to be shipped to a customer. The shipping company charges by weight and volume (in cubic feet) in shipping containers.

_____ **1.** How many lamps are on a pallet?

_____ **2.** What is the total weight of all the lamps on a pallet?

_____ **3.** Using the pallet dimensions, what is the volume (in cubic feet) of the pallet of lamps?

Electrical Power Generation Objectives

- Convert a listed value for power generated from one metric prefix value to another metric prefix value.
- Convert a listed value for power generated from a metric number to a base number.

Procedure 6-4

Click on the Electrical Power Data Sheets button on the CD-ROM. Next, click on the Total Electric Power Industry Summary Statistics button and answer the following questions. Although wind-generated power increased, conventional hydroelectric energy production actually decreased by 2,618,000 megawatt-hours (2,618,000,000,000 watt-hours) in August 2011 compared to August 2012.

_____ **1.** How many gigawatt-hours (GWh) was this reduction?

For all sectors total, nuclear power generated 69,602,000,000,000 megawatt-hours of power in August 2012.

_____ **2.** This number in its equivalent base units is ___.

Numbering Systems and Codes

Review Questions

7

Name _____ Date _____

True-False

T   F   **1.** Prefixes and exponents may be used to simplify the expression of large decimal numbers.

T   F   **2.** The United States monetary system is based on the octal numbering system.

T   F   **3.** A word is a group of one or more bytes (1111000011110000, etc.).

T   F   **4.** The binary coded decimal (BCD) system is a coding system that represents each digit from zero through nine as an eight-bit binary number.

T   F   **5.** The first two color bands represent the first two digits in the value of a resistor.

T   F   **6.** Capacitor color codes are standardized by the Electronic Industries Association (EIA).

T   F   **7.** The fourth band indicates how far the actual measured value of a resistor can be from the coded value.

T   F   **8.** Digital electronic systems use the binary numbering system because it uses the digits 0 and 1, which are used to represent the two physical states that most electrical elements have (ON/OFF, open/closed, etc.).

Multiple Choice

_____ **1.** A(n) ___ number is a number expressed in base 10.
    A. octal
    B. hexadecimal
    C. decimal
    D. binary

_____ **2.** A ___ is each digit (0 or 1) of a binary number.
    A. bit
    B. byte
    C. nibble
    D. word

_____ **3.** The ___ numbering system is a base 16 numbering system that uses digits 0, 1, 2, 3, 4, 5, 6, 7, 8, 9, and the first six letters of the alphabet, A, B, C, D, E, and F.
  A. decimal
  B. binary
  C. octal
  D. hexadecimal

_____ **4.** Color-coded capacitance values are always in ___ units.
  A. picofarad
  B. microfarad
  C. millifarad
  D. farad

_____ **5.** As many as ___ digits could change for a single count when using the binary number system.
  A. 2
  B. 4
  C. 6
  D. 8

Completion

_____ **1.** The ___ numbering system is the most commonly used numbering system.

_____ **2.** The ___ numbering system is a base two numbering system that uses only two digits, 0 and 1.

_____ **3.** A(n) ___ is a group of 4 bits (1001, 0011, etc.).

_____ **4.** A(n) ___ is a group of 8 bits (10001101, etc.).

_____ **5.** The ___ numbering system is a base eight numbering system that uses only digits 0, 1, 2, 3, 4, 5, 6, and 7.

_____ **6.** Mica capacitors normally use a(n) ___ dot system.

_____ **7.** In the ___ code, only one digit changes for each count from one number to the next.

_____ **8.** For tubular ceramic capacitors, the wide color band specifying temperature coefficient indicates the ___ end.

_____ **9.** ___ are devices that limit the current flowing in an electronic circuit.

_____ **10.** To convert a decimal number to a(n) ___ number, the decimal number is divided by 2 in repetitive steps.

Numbering Systems and Codes

Worksheet 7-1

Name _____ Date _____

Binary/Decimal and Octal/Decimal Conversion

_____ 1. Binary 2^{14} = decimal ___

_____ 2. Binary 2^{12} = decimal ___

_____ 3. Binary 2^5 = decimal ___

_____ 4. Binary 2^4 = decimal ___

_____ 5. Binary 2^3 = decimal ___

_____ 6. Binary 2^2 = decimal ___

_____ 7. Binary 2^1 = decimal ___

_____ 8. Decimal 1 = binary ___

_____ 9. Decimal 2 = binary ___

_____ 10. Decimal 3 = binary ___

_____ 11. Decimal 4 = binary ___

_____ 12. Decimal 10 = binary ___

_____ 13. Decimal 14 = binary ___

_____ 14. Decimal 15 = binary ___

_____ 15. Binary 0000100101 = decimal ___

_____ 16. Binary 0010000001 = decimal ___

_____ 17. Binary 0110000110 = decimal ___

_____ 18. Binary 1110000100 = decimal ___

_____ 19. Decimal 450 = binary ___

_____ 20. Decimal 1305 = binary ___

_____ 21. Octal 8^{10} = decimal ___

_____ 22. Octal 8^9 = decimal ___

_____ 23. Octal 8^8 = decimal ___

_____ 24. Octal 8^7 = decimal ___

_____ 25. Octal 8^3 = decimal ___

_____ 26. Octal 8^2 = decimal ___

_____ 27. Octal 8^1 = decimal ___

_____ 28. Decimal 1 = octal ___

_____ 29. Decimal 2 = octal ___

_____ 30. Decimal 6 = octal ___

_____ 31. Decimal 7 = octal ___

_____ 32. Decimal 8 = octal ___

_____ 33. Decimal 10 = octal ___

_____ 34. Decimal 14 = octal ___

_____ 35. Decimal 15 = octal ___

_____ 36. Decimal 20 = octal ___

_____ 37. Octal 121 = decimal ___

_____ 38. Octal 795 = decimal ___

_____ 39. Octal 2273 = decimal ___

_____ 40. Decimal 64 = octal ___

_____ 41. Decimal 331 = octal ___

_____ 42. Decimal 3186 = octal ___

Numbering Systems and Codes

Worksheet 7-2

Name _____ **Date** _____

Hexadecimal/Decimal Conversion

_____ 1. Hexadecimal 16^8 = decimal ___

_____ 2. Hexadecimal 16^7 = decimal ___

_____ 3. Hexadecimal 16^6 = decimal ___

_____ 4. Hexadecimal 16^5 = decimal ___

_____ 5. Hexadecimal 16^4 = decimal ___

_____ 6. Hexadecimal 16^3 = decimal ___

_____ 7. Hexadecimal 16^2 = decimal ___

_____ 8. Hexadecimal 16^1 = decimal ___

_____ 9. Hexadecimal 16^0 = decimal ___

_____ 10. Hexadecimal A1 = decimal ___

_____ 11. Hexadecimal C4 = decimal ___

_____ 12. Hexadecimal F6 = decimal ___

_____ 13. Hexadecimal B32 = decimal ___

_____ 14. Hexadecimal E55 = decimal ___

_____ 15. Hexadecimal D523 = decimal ___

_____ 16. Hexadecimal A152 = decimal ___

_____ 17. Hexadecimal B6011 = decimal ___

_____ 18. Hexadecimal E45002 = decimal ___

_____ 19. Hexadecimal A245 = decimal ___

_____ 20. Decimal 3602 = hexadecimal ___

_____ 21. Decimal 49,409 = hexadecimal ___

_____ 22. Decimal 2677 = hexadecimal ___

_____ 23. Decimal 57,942 = hexadecimal ___

_____ 24. Decimal 2882 = hexadecimal ___

_____ 25. Decimal 213 = hexadecimal ___

_____ 26. Decimal 180 = hexadecimal ___

_____ 27. Decimal 167 = hexadecimal ___

_____ 28. Decimal 3360 = hexadecimal ___

Name _____ **Date** _____

Resistor Color Codes

_____ **1.** Resistance = ___ Ω

_____ **2.** Tolerance = ±___%

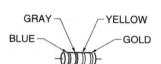

BLACK — — RED
ORANGE — — GOLD

_____ **3.** Resistance = ___ Ω

_____ **4.** Tolerance = ±___%

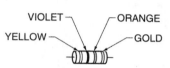

VIOLET — — ORANGE
YELLOW — — GOLD

_____ **5.** Resistance = ___ Ω

_____ **6.** Tolerance = ±___%

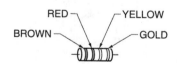

GRAY — — YELLOW
BLUE — — GOLD

_____ **7.** Resistance = ___ Ω

_____ **8.** Tolerance = ±___%

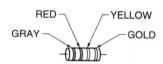

RED — — YELLOW
GRAY — — GOLD

_____ **9.** Resistance = ___ Ω

_____ **10.** Tolerance = ±___%

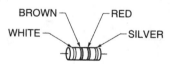

RED — — YELLOW
BROWN — — GOLD

_____ **11.** Resistance = ___ Ω

_____ **12.** Tolerance = ±___%

BROWN — — RED
WHITE — — SILVER

_____ **13.** Resistance = ___ Ω

_____ **14.** Tolerance = ±___%

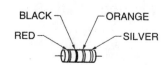

BLACK — — ORANGE
RED — — SILVER

_____ **15.** Resistance = ___ Ω

_____ **16.** Tolerance = ±___%

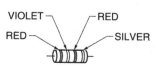

VIOLET — — RED
RED — — SILVER

69

Numbering Systems and Codes

Worksheet 7-4

Name _____ **Date** _____

Capacitor Color Codes

_____ **1.** Capacitance = ___ pF

_____ **2.** Tolerance = ±___%

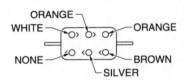

ORANGE — ORANGE
WHITE —
NONE —
— BROWN
— SILVER

_____ **3.** Capacitance = ___ pF

_____ **4.** Tolerance = ±___%

BLUE —
WHITE — — GRAY
NONE — — BROWN
— GREEN

_____ **5.** Capacitance = ___ pF

_____ **6.** Tolerance = ±___%

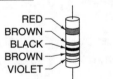

ORANGE —
BLACK — — WHITE
NONE — — BROWN
— GOLD

_____ **7.** Capacitance = ___ pF

_____ **8.** Tolerance = ±___%

GRAY —
WHITE — — RED
NONE — — BLACK
— SILVER

_____ **9.** Capacitance = ___ pF

_____ **10.** Tolerance = ±___%

RED —
BROWN —
BLACK —
BROWN —
VIOLET —

_____ **11.** Capacitance = ___ pF

_____ **12.** Tolerance = ±___%

RED —
RED —
VIOLET —
BROWN —
WHITE —

_____ **13.** Capacitance = ___ pF

_____ **14.** Tolerance = ±___%

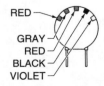

RED —
GRAY —
RED —
BLACK —
VIOLET —

_____ **15.** Capacitance = ___ pF

_____ **16.** Tolerance = ±___%

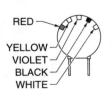

RED —
YELLOW —
VIOLET —
BLACK —
WHITE —

© 2013 American Technical Publishers, Inc.
All rights reserved

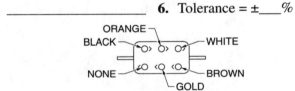

Numbering Systems and Codes

Activities

7

Name _____ Date _____

Hands-On Activity Objectives

- Determine the numerical resistance value for each resistor based on its color code.
- Determine the tolerance rating for each resistor based on its color code.

Procedure 7-1

Determine the resistor numerical value and percentage rating.

_____ **1.** The numerical value of a resistor with a color code of red, black, black, and silver is ___ Ω.

_____ **2.** The tolerance rating of a resistor with a color code of red, black, black, and silver is ±___%.

_____ **3.** The numerical value of a resistor with a color code of violet, green, black, and gold is ___ Ω.

_____ **4.** The tolerance rating of a resistor with a color code of violet, green, black, and gold is ±___%.

_____ **5.** The numerical value of a resistor with a color code of brown, black, brown, and none is ___ Ω.

_____ **6.** The tolerance rating of a resistor with a color code of brown, black, brown, and none is ±___%.

_____ **7.** The numerical value of a resistor with a color code of brown, green, brown, and silver is ___ Ω.

_____ **8.** The tolerance rating of a resistor with a color code of brown, green, brown, and silver is ±___%.

_____ **9.** The numerical value of a resistor with a color code of red, black, brown, and gold is ___ Ω.

_____ **10.** The tolerance rating of a resistor with a color code of red, black, brown, and gold is ±___%.

_____ **11.** The numerical value of a resistor with a color code of green, brown, brown, and silver is ___ Ω.

_____ **12.** The tolerance rating of a resistor with a color code of green, brown, brown, and silver is ±___%.

_____ **13.** The numerical value of a resistor with a color code of brown, black, red, and none is ___ Ω.

_____ **14.** The tolerance rating of a resistor with a color code of brown, black, red, and none is ±___%.

_____ **15.** The numerical value of a resistor with a color code of green, blue, red, and silver is ___ Ω.

_____ **16.** The tolerance rating of a resistor with a color code of green, blue, red, and silver is ±___%.

Procedure 7-2

Using the resistance values and tolerance ratings from Procedure 7-1, calculate the resistance range for each resistor.

_____ 1. The resistance range of a resistor with a color code of red, black, black, and silver is ___.

_____ 2. The resistance range of a resistor with a color code of violet, green, black, and gold is ___.

_____ 3. The resistance range of a resistor with a color code of brown, black, brown, and none is ___.

_____ 4. The resistance range of a resistor with a color code of brown, green, brown, and silver is ___.

_____ 5. The resistance range of a resistor with a color code of red, black, brown, and gold is ___.

_____ 6. The resistance range of a resistor with a color code of green, brown, brown, and silver is ___.

Measurement Activity Objectives

• Apply the procedure for measuring resistance using a DMM.
• Measure the resistance value of each resistor to verify that the actual value is within its tolerance range.

Procedure 7-3

Measure the resistance of each resistor by applying the Ohmmeter—Resistance Measurement procedure in the Appendix. Determine if each resistor is within its tolerance range.

_____ 1. The measured resistance of the resistor with a color code of red, black, black, and gold is ___ Ω.

_____ 2. Is the resistor within its tolerance range?

_____ 3. The measured resistance of the resistor with a color code of violet, green, black, and gold is ___ Ω.

_____ 4. Is the resistor within its tolerance range?

_____ 5. The measured resistance of the resistor with a color code of brown, black, brown, and gold is ___ Ω.

_____ 6. Is the resistor within its tolerance range?

_____ 7. The measured resistance range of the resistor with a color code of brown, green, brown, and gold is ___ Ω.

_____ 8. Is the resistor within its tolerance range?

_____ 9. The measured resistance range of the resistor with a color code of red, black, brown, and gold is ___ Ω.

_____ 10. Is the resistor within its tolerance range?

_____ 11. The measured resistance range of the resistor with a color code of green, brown, brown, and gold is ___ Ω.

_____ 12. Is the resistor within its tolerance range?

Meter Abbreviations and Displays

Review Questions

8

Name _____ Date _____

True-False

T   F   **1.** A permanent meter is used to take momentary measurements.

T   F   **2.** Abbreviations are independent of language because they can be recognized regardless of the language a person speaks.

T   F   **3.** An analog display is an electromechanical device that indicates readings by the mechanical motion of a pointer.

T   F   **4.** A range switch determines the placement of the decimal point.

T   F   **5.** Any hand-held DMM should be able to take basic voltage, DC current, and resistance measurements.

Multiple Choice

_____ **1.** A(n) ___ is a graph composed of segments that function as an analog pointer.
- A. analog display
- B. bar graph
- C. digital display
- D. linear graph

_____ **2.** A(n) ___ voltage is voltage that appears on a meter that is not connected to a circuit.
- A. magnetic
- B. infrared
- C. ghost
- D. vector

_____ **3.** A(n) ___ bar graph is a bar graph that displays a fraction of the full range on the graph.
- A. closed
- B. fractional
- C. open
- D. wrap-around

73

_____ **4.** A bar graph reading is updated ___ times per second.
  A. 10
  B. 20
  C. 30
  D. 40

_____ **5.** A ___ is a division that divides secondary divisions in halves, thirds, fourths, fifths, etc.
  A. primary division
  B. subdivision
  C. sublevel
  D. fractional division

Completion

_____ **1.** A(n) ___ meter is a meter capable of measuring and displaying only one quantity.

_____ **2.** A(n) ___ is a meter that is capable of measuring two or more quantities.

_____ **3.** A(n) ___ meter is installed to constantly measure and display quantities.

_____ **4.** A(n) ___ is a letter or combination of letters that represents a word.

_____ **5.** A(n) ___ is a graphic element that represents a quantity or unit.

_____ **6.** A(n) ___ scale is a scale that is divided into equally spaced segments.

_____ **7.** A(n) ___ scale is a scale that is divided into unequally spaced segments.

_____ **8.** A(n) ___ display is an electronic device that displays readings as numerical values.

_____ **9.** A(n) ___ division is a division that divides primary divisions in halves, thirds, fourths, fifths, etc.

_____ **10.** ___ are produced by magnetic fields generated by current-carrying conductors, fluorescent lighting, and operating electrical equipment.

Reading Analog Displays

_____ **1.** Primary division

_____ **2.** Secondary division

_____ **3.** Subdivision

Name _____ **Date** _____

Reading Analog Displays

_____ 1. Meter 1 reading = ____ V with range switch set on 1 V.

_____ 2. Meter 1 reading = ____ V with range switch set on 10 V.

_____ 3. Meter 1 reading = ____ V with range switch set on 100 V.

_____ 4. Meter 2 reading = ____ Ω with range switch set on R × 1.

_____ 5. Meter 2 reading = ____ Ω with range switch set on R × 100.

_____ 6. Meter 2 reading = ____ Ω with range switch set on R × 1k.

_____ 7. Meter 3 reading = ____ V with range switch set on 1 V.

_____ 8. Meter 3 reading = ____ V with range switch set on 10 V.

_____ 9. Meter 3 reading = ____ V with range switch set on 100 V.

_____ 10. Meter 4 reading = ____ V with range switch set on 1 V.

_____ 11. Meter 4 reading = ____ V with range switch set on 1 V.

_____ 12. Meter 4 reading = ____ V with range switch set on 1 V.

_____ 13. Meter 5 reading = ____ Ω with range switch set on R × 1.

_____ 14. Meter 5 reading = ____ Ω with range switch set on R × 100.

_____ 15. Meter 5 reading = ____ Ω with range switch set on R × 1k.

METER 1

METER 2

METER 3

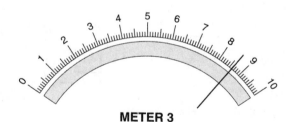

METER 4

METER 5

75

Name _____ **Date** _____

Reading Digital Displays

Determine the resolution (0.001 V, 0.01 V, 0.1 V, or 1 V) for each meter reading.

_____ **1.** Resolution = ___ V

| 1.02 ᵛ |

_____ **2.** Resolution = ___ V

| 480 ᵛ |

_____ **3.** Resolution = ___ V

| 175 ᵛ |

_____ **4.** Resolution = ___ V

| 480.3 ᵛ |

_____ **5.** Resolution = ___ V

| 208.3 ᵛ |

_____ **6.** Resolution = ___ V

| 20.22 ᵛ |

_____ **7.** Resolution = ___ V

| 0.017 ᵛ |

_____ **8.** Resolution = ___ V

| 1 ᵛ |

Reading Meter Voltages

_____ **1.** Voltage reading = ___ mV

_____ **2.** Each segment = ___ mV

RANGE SWITCH
SET ON 30 mV

_____ **3.** Voltage reading = ___ V

_____ **4.** Each segment = ___ V

RANGE SWITCH
SET ON 30 V

_____ **5.** Voltage reading = ___ V

_____ **6.** Each segment = ___ V

RANGE SWITCH
SET ON 300 V

_____ **7.** Voltage reading = ___ mV

_____ **8.** Each segment = ___ mV

RANGE SWITCH
SET ON 300 mV

Meter Abbreviations and Displays

Activities

8

Name _____ Date _____

Virtual Meter Activity Objectives

- Identify abbreviations on DMMs used to simplify electrical terms (voltage, current, etc.) and measurements.
- Learn the meaning of each meter abbreviation and symbol used on DMMs.

Procedure 8-1

Click on the Virtual Meters button on the CD-ROM. Next, click on the Virtual DMM 87V Demo button. Set the DMM function switch to VAC.

_____ **1.** What abbreviations appear on the meter display?

Set the DMM function switch to VDC.

_____ **2.** What new abbreviation appears that was not on the last display?

Set the DMM function switch to mVDC.

_____ **3.** What new abbreviation appears that was not on the last display?

With the DMM still on the mVDC setting, press the orange function button.

_____ **4.** What new abbreviation appears?

With the DMM still on the mVDC setting, press the range function button.

_____ **5.** What new abbreviation appears?

Set the DMM function switch to the resistance measurement setting.

_____ **6.** What new symbol appears?

With the DMM still on the resistance measurement setting, press the yellow function button.

_____ **7.** What new abbreviation appears?

Set the DMM to the diode test position.

_____ **8.** What is a typical forward-biased voltage measurement on a good diode?

Set the DMM to the mA/A measuring position.

_____ **9.** What happens to the positive lead of the meter?

_____ **10.** With the DMM still set to the mA/A position, what does pressing the orange function button do?

Measurement Activity Objectives

- Determine the circuit parameters (VAC, VDC, etc.) that can be measured with a given DMM.
- Determine the meter features (recording minimum/maximum values, etc.) included on a given DMM.

Procedure 8-2

Obtain a DMM. Check all the functions that the DMM has available.

YES NO

☐ ☐ **1.** AC voltage measurement

☐ ☐ **2.** DC voltage measurement

☐ ☐ **3.** DC millivolt measurement

☐ ☐ **4.** Resistance measurement

☐ ☐ **5.** AC milliamp measurement

☐ ☐ **6.** DC milliamp measurement

☐ ☐ **7.** AC amp measurement

☐ ☐ **8.** DC amp measurement

☐ ☐ **9.** AC microamp measurement

☐ ☐ **10.** DC microamp measurement

☐ ☐ **11.** Diode check function

☐ ☐ **12.** Temperature measurement function

☐ ☐ **13.** Capacitance measurement function

14. List other functions available.

Name _____ **Date** _____

True-False

T F **1.** An ammeter is used to measure the amount of voltage in a circuit that is powered.

T F **2.** For most electrical circuits, the voltage level may vary ±10% from the rated voltage.

T F **3.** Amperage measurements are normally taken to indicate the amount of circuit loading or the condition of a load.

T F **4.** Many meters include a fuse in the low-ampere range to prevent meter damage caused by excessive current.

T F **5.** High voltage applied to a meter set to measure resistance causes meter damage.

T F **6.** Megohmmeter test voltages range from 50 V to 5000 V.

T F **7.** A receptacle tester is a test instrument that indicates the presence of voltage when the test tip touches, or is near, an energized hot conductor or energized metal part.

T F **8.** To measure frequency, the scope probes are connected in parallel with the circuit or component under test.

Multiple Choice

_____ **1.** A ___ tachometer is a device that measures the speed of an object without direct contact with the object.
   A. contact
   B. photo
   C. linear
   D. hand-held

_____ **2.** A(n) ___ is a special DC voltmeter that detects the presence or absence of a signal.
   A. multimeter
   B. digital logic probe
   C. scopemeter
   D. ohmmeter

79

_____ **3.** ___ is the amount of electrical pressure in a circuit.
      A. Voltage
      B. Current
      C. Resistance
      D. Temperature

_____ **4.** ___ is the measurement of the intensity of heat.
      A. Voltage
      B. Temperature
      C. Resistance
      D. Current

_____ **5.** Digital circuits fail because the ___ is lost somewhere between the circuit input and output stages.
      A. temperature
      B. signal
      C. resistance
      D. voltage

Completion

_____ **1.** A(n) ___ is used to measure the amount of current flowing in a circuit that is powered.

_____ **2.** A(n) ___ ammeter is a meter that measures current in a circuit by inserting the meter in series with the component(s) under test.

_____ **3.** A(n) ___ ammeter is a meter that measures current in a circuit by measuring the strength of the magnetic field around a conductor.

_____ **4.** A(n) ___ is a device that is used to measure the amount of resistance in a component (or circuit) that is not powered.

_____ **5.** A(n) ___ is a test instrument with a bulb that is connected to two test leads to give a visual indication when voltage is present in a circuit.

_____ **6.** ___ ammeters are used to measure currents from 0.01 A or less to 1000 A or more.

_____ **7.** A(n) ___ is a device that detects insulation deterioration by measuring high resistance values under high test voltage conditions.

_____ **8.** A(n) ___ is an instrument that measures temperature at a single point.

_____ **9.** A(n) ___ tachometer is a device that measures the rotational speed of an object through direct contact of the tachometer tip with the object to be measured.

_____ **10.** A(n) ___ is a device that detects heat patterns in the infrared-wavelength spectrum without making direct contact with equipment.

_____ **11.** A(n) ___ tachometer is a device that uses a flashing light to measure the speed of a moving object.

_____ **12.** A(n) ___ is a device that gives a visual display of voltages.

_____ **13.** A(n) ___ is a reference point/line that is visually displayed on the face of the scope screen.

_____ **14.** A(n) ___ is an instrument that displays an instantaneous voltage.

_____ **15.** ___ is the movement of the displayed trace across the scope screen.

Meters

_____ **1.** Measures resistance

_____ **2.** Measures DC voltage

_____ **3.** Measures high resistance values

_____ **4.** Measures AC voltage

_____ **5.** Measures speed

_____ **6.** Measures current

A   **B**   **C**   **D**   **E**   **F**

Name _____ **Date** _____

AC Voltage Measurements

1. Draw the correct position of the function switch and test leads to measure the voltage at the receiver outlet.

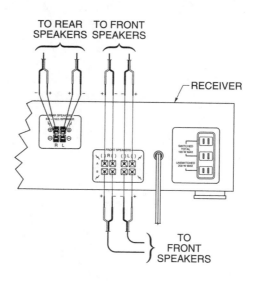

2. Draw the correct position of the function switch and leads to measure the voltage out of the transformer.

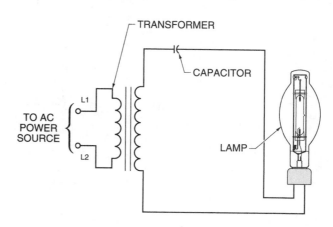

Taking Standard Measurements

Worksheet 9-2

Name _____ **Date** _____

DC Voltage Measurements

1. Draw the correct position of the function switch and red and black test leads to measure the voltage out of the thermocouple. Label the red and black test leads.

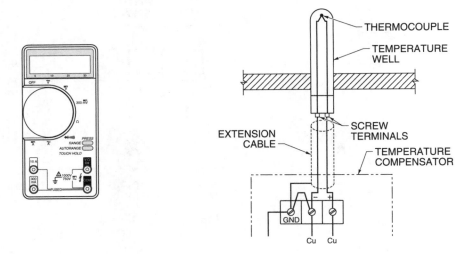

2. Draw the correct position of the function switch and red and black test leads to measure the voltage at the battery. Label the red and black test leads.

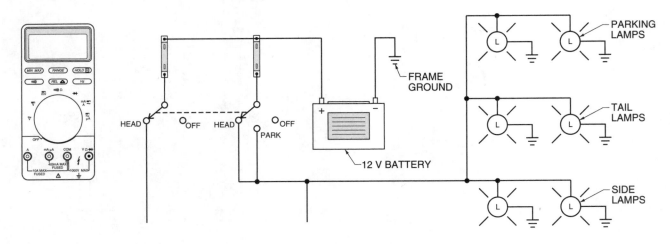

Name _____ **Date** _____

In-Line AC Current Measurements

1. Draw the correct position of the function switch and test leads to measure the current out of the bridge rectifier circuit.

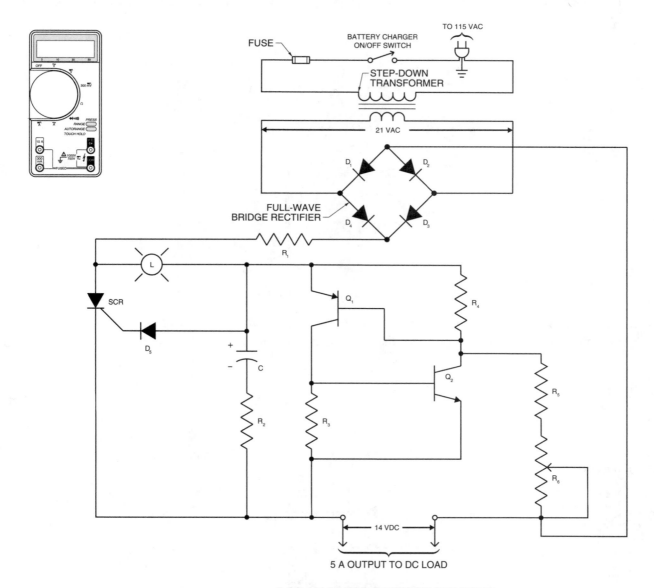

5 A/14 V RATED BATTERY CHARGER

Taking Standard Measurements

Worksheet 9-4

Name _____ Date _____

Clamp-On AC/DC Current Measurements

1. Draw the correct position of the function switch to measure the AC current at the main power panel neutral wire.

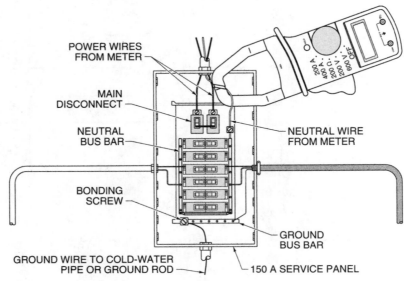

POWER WIRES FROM METER

MAIN DISCONNECT

NEUTRAL BUS BAR

NEUTRAL WIRE FROM METER

BONDING SCREW

GROUND BUS BAR

GROUND WIRE TO COLD-WATER PIPE OR GROUND ROD

150 A SERVICE PANEL

2. Draw the correct position of the function switch and red and black test leads to measure the voltage out of the DC side of the alternator. Label the red and black test leads.

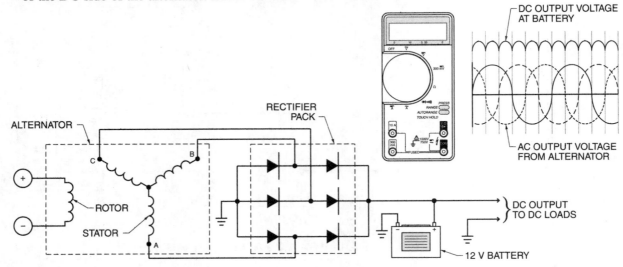

ALTERNATOR

RECTIFIER PACK

DC OUTPUT VOLTAGE AT BATTERY

AC OUTPUT VOLTAGE FROM ALTERNATOR

ROTOR

STATOR

DC OUTPUT TO DC LOADS

12 V BATTERY

Taking Standard Measurements

Worksheet 9-5

Name _____ Date _____

Resistance Measurements

1. Draw the correct position of the function switch and test leads to measure the resistance out of the photocell.

2. Draw the correct position of the function switch and test leads to measure the resistance of the electrical contacts.

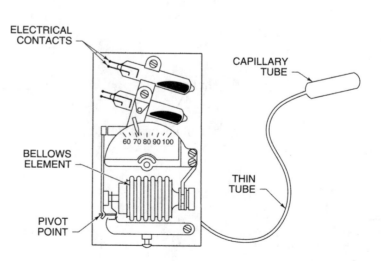

Taking Standard Measurements

Worksheet 9-6

Name _____ Date _____

Oscilloscope/Scopemeter Measurements

_____ **1.** Peak-to-peak voltage = ___ V

_____ **2.** rms voltage = ___ V

_____ **3.** Frequency = ___ Hz

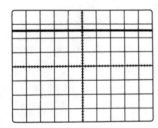
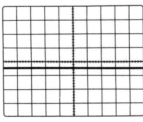

_____ **4.** Voltage = ___ V

_____ **5.** Voltage = ___ V

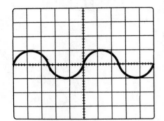
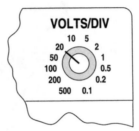
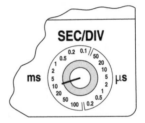

_____ **6.** Minimum peak voltage = ___ V.

_____ **8.** Minimum peak voltage = ___ V.

_____ **7.** Maximum peak voltage = ___ V.

_____ **9.** Maximum peak voltage = ___ V.

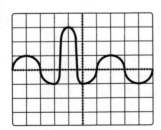
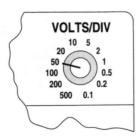
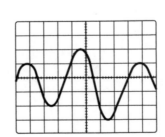
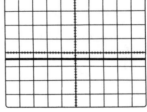

Taking Standard Measurements

Activities

9

Name _____ **Date** _____

Virtual Meter Activity Objectives

- Set a digital meter to measure VAC, VDC, mVDC, resistance, capacitance, diodes, A/mA/µA AC, and A/mA/µA DC.
- Convert displayed meter readings that include a prefix to a base number value.

Procedure 9-1

Click on the Virtual Meters button on the CD-ROM. Next, click on the Virtual DMM 87V Demo button. Move the position of the selector switch to the VAC position.

1. What is the displayed reading?

Move the position of the selector switch to the VDC position.

2. What is the displayed reading?

Move the position of the selector switch to the mVDC position.

3. What is the displayed reading?

Move the position of the selector switch to the ohm (Ω) position.

4. What is the displayed reading?

89

From the ohm (Ω) position, press the orange button to set the meter to a capacitance measuring mode.

5. What is the displayed value in μF?

6. What is the displayed value in mF?

7. What is the displayed value in F?

Move the position of the selector switch to the diode test position.

8. Is the displayed value within the acceptable range for a silicon junction diode?

Move the position of the selector switch to the AAC position.

9. What is the displayed reading?

Use the orange button to switch to an ADC measurement.

10. What is the displayed reading?

Move the position of the selector switch to the μA AC position.

11. What is the displayed reading?

Change the meter to take an μA DC measurement.

12. What is the displayed reading?

Name _____ **Date** _____

True-False

T   F   **1.** Heat power sources such as thermocouples convert heat energy into electrical energy.

T   F   **2.** Contacts are the conducting part of a switch that operates with another conducting part to make or break a circuit.

T   F   **3.** A power source is a device that converts various forms of energy into electricity.

T   F   **4.** A power source may be chemical (battery), magnetic (generator), heat (thermocouple), or solar (photovoltaic cell).

T   F   **5.** Cells are connected in series to increase current or are connected in parallel to increase voltage.

T   F   **6.** Solar power sources such as photoconductive cells convert heat energy into electricity.

T   F   **7.** A capacitor is a device that changes resistance with a change in temperature.

T   F   **8.** A proximity switch is a switch that detects the presence or absence of an object without touching the object.

Multiple Choice

_____ **1.** A ___ is a device that converts fluid power to mechanical power.
  A. fluid actuator
  B. check valve
  C. lever actuator
  D. flow control valve

_____ **2.** A ___ is a device used to allow the fluid in a fluid power system to travel in only one direction.
  A. lever actuator
  B. check valve
  C. fluid actuator
  D. flow control valve

91

_____ **3.** A ___ actuator is an actuator that automatically returns a valve to a set position.
    A. palm
    B. foot
    C. detent
    D. spring

_____ **4.** A(n) ___ is a device that reduces the noise that is made when air passes through a restriction.
    A. manual shutoff valve
    B. air muffler
    C. accumulator
    D. pressure gauge

_____ **5.** A ___ valve is a valve that limits the amount of pressure in a fluid power system.
    A. sequence
    B. pressure-reducing
    C. manual shutoff
    D. pressure relief

Completion

_____ **1.** A(n) ___ is a letter or combination of letters that represents a word.

_____ **2.** A(n) ___ is a word formed from the first letter(s) of a compound term.

_____ **3.** A(n) ___ is a graphic element that represents an operation, a quantity, or a unit.

_____ **4.** A(n) ___ is a switch that disconnects electrical circuits from motors and machines.

_____ **5.** A(n) ___ is a switch that makes (NO) or breaks (NC) a circuit when manually pressed.

_____ **6.** Electrical power such as alternating current is supplied by a(n) ___.

_____ **7.** A(n) ___ switch is a switch with an operator that is rotated to activate the electrical contacts.

_____ **8.** A(n) ___ is a control device that uses a preset time period as part of the control function.

_____ **9.** A(n) ___ switch is a switch that converts mechanical motion into an electrical signal.

_____ **10.** A(n) ___ is an electric output device that converts electrical energy into linear mechanical force.

_____ **11.** A(n) ___ is a solid-state switching device that switches current ON by a quick pulse of control current.

_____ **12.** A(n) ___ is an electrically-operated switch (contactor) that includes motor overload protection.

_____ **13.** A(n) ___ is a semiconductor device that offers very high opposition to current flow in one direction and very low opposition to current flow in the opposite direction.

_____ **14.** A(n) ___ is a winding consisting of insulated conductors arranged to produce magnetic flux.

_____ **15.** A(n) ___ is an electrical interface designed to change AC from one voltage level to another voltage level.

Symbols

_____ **1.** Temperature switch

_____ **2.** Thermal overload

_____ **3.** Oscillator

_____ **4.** Bell

_____ **5.** Fuse

_____ **6.** Capacitor

_____ **7.** Battery

_____ **8.** Normally open contacts

_____ **9.** Solenoid

_____ **10.** NPN transistor

Symbols and Printreading

Worksheet 10-1

10

Name _____ **Date** _____

Abbreviations

_____ 1. Abbreviation at A = ____

_____ 2. Abbreviation at B = ____

_____ 3. Abbreviation at C = ____

_____ 4. Abbreviation at D = ____

_____ 5. Abbreviation at E = ____

_____ 6. Abbreviation at F = ____

_____ 7. Abbreviation at G = ____

_____ 8. Abbreviation at H = ____

_____ 9. Abbreviation at I = ____

_____ 10. Abbreviation at J = ____

_____ 11. Abbreviation at K = ____

_____ 12. Abbreviation at L = ____

_____ 13. Abbreviation at M = ____

_____ 14. Abbreviation at N = ____

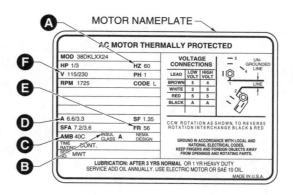

MOTOR NAMEPLATE

AC MOTOR THERMALLY PROTECTED

MOD 38DKLXX24		VOLTAGE CONNECTIONS		
HP 1/3	HZ 60			
V 115/230	PH 1	LEAD	LOW VOLT	HIGH VOLT
RPM 1725	CODE L	BROWN	5	4
		WHITE	2	5
		RED	5	5
		BLACK	A	A
A 6.6/3.3	SF 1.35			
SFA 7.2/3.6	FR 56			
AMB 40C	INSUL CLASS A	NEMA DESIGN		
TIME RATING CONT.				
SER. NO. MWT				

CCW ROTATION AS SHOWN, TO REVERSE ROTATION INTERCHANGE BLACK & RED

GROUND IN ACCORDANCE WITH LOCAL AND NATIONAL ELECTRICAL CODES. KEEP FINGERS AND FOREIGN OBJECTS AWAY FROM OPENINGS AND ROTATING PARTS.

LUBRICATION: AFTER 3 YRS NORMAL OR 1 YR HEAVY DUTY SERVICE ADD OIL ANNUALLY. USE ELECTRIC MOTOR OR SAE 10 OIL.

MADE IN U.S.A.

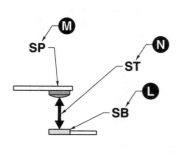

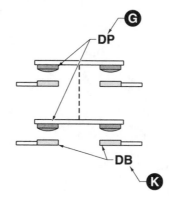

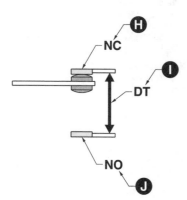

© 2013 American Technical Publishers, Inc.
All rights reserved

Name _____ **Date** _____

Acronyms

_____ **1.** Acronym at A = ___

_____ **2.** Acronym at B = ___

_____ **3.** Acronym at C = ___

_____ **4.** Acronym at D = ___

_____ **5.** Acronym at E = ___

_____ **6.** Acronym at F = ___

_____ **7.** Acronym at G = ___

_____ **8.** Acronym at H = ___

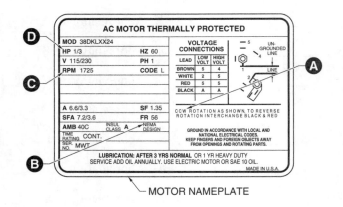

AC MOTOR THERMALLY PROTECTED

— MOTOR NAMEPLATE

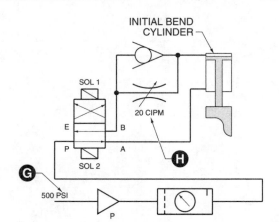

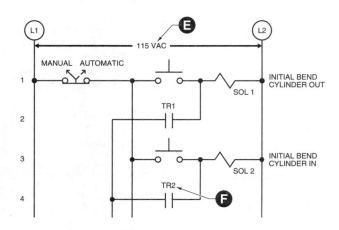

Name _____ Date _____

Electrical Symbols

_____ **1.** Symbol at A = ___

_____ **2.** Symbol at B = ___

_____ **3.** Symbol at C = ___

_____ **4.** Symbol at D = ___

_____ **5.** Symbol at E = ___

_____ **6.** Symbol at F = ___

_____ **7.** Symbol at G = ___

_____ **8.** Symbol at H = ___

_____ **9.** Symbol at I = ___

_____ **10.** Symbol at J = ___

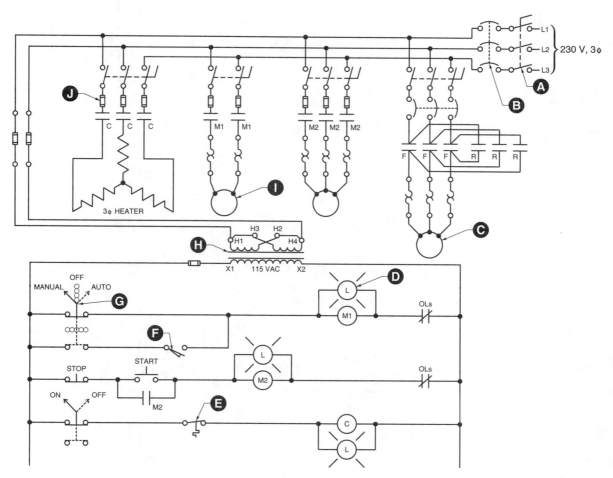

Name _____ Date _____

Electronic Symbols

_____ **1.** Symbol at A = ___

_____ **2.** Symbol at B = ___

_____ **3.** Symbol at C = ___

_____ **4.** Symbol at D = ___

_____ **5.** Symbol at E = ___

_____ **6.** Symbol at F = ___

_____ **7.** Symbol at G = ___

_____ **8.** Symbol at H = ___

_____ **9.** Symbol at I = ___

_____ **10.** Symbol at J = ___

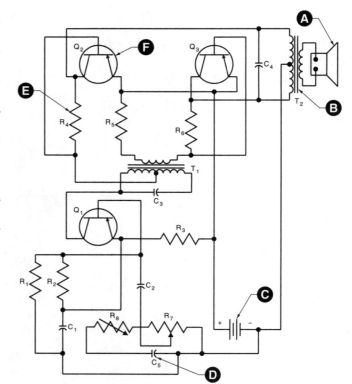

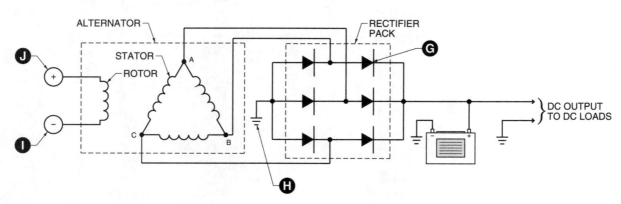

Name _____ Date _____

Fluid Power Symbols

_____ 1. Symbol at A = ___

_____ 2. Symbol at B = ___

_____ 3. Symbol at C = ___

_____ 4. Symbol at D = ___

_____ 5. Symbol at E = ___

_____ 6. Symbol at F = ___

_____ 7. Symbol at G = ___

_____ 8. Symbol at H = ___

_____ 9. Symbol at I = ___

_____ 10. Symbol at J = ___

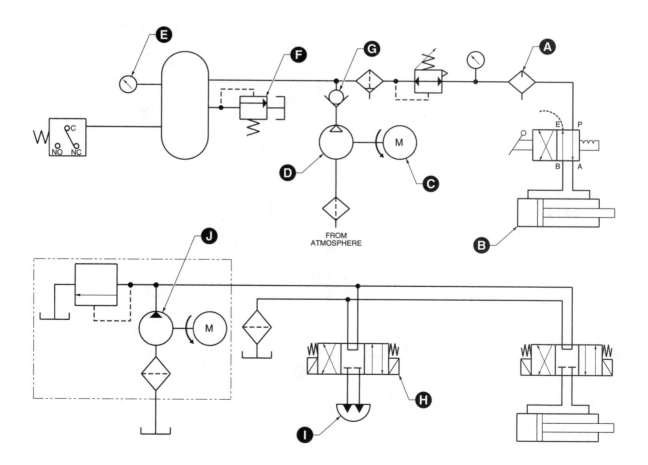

Symbols and Printreading 10

Worksheet 10-6

Name ______________________ **Date** ______________

Component Usage

______ **1.** Component at A use = ___

______ **2.** Component at B use = ___

______ **3.** Component at C use = ___

______ **4.** Component at D use = ___

______ **5.** Component at E use = ___

______ **6.** Component at F use = ___

______ **7.** Component at G use = ___

______ **8.** Component at H use = ___

______ **9.** Component at I use = ___

______ **10.** Component at J use = ___

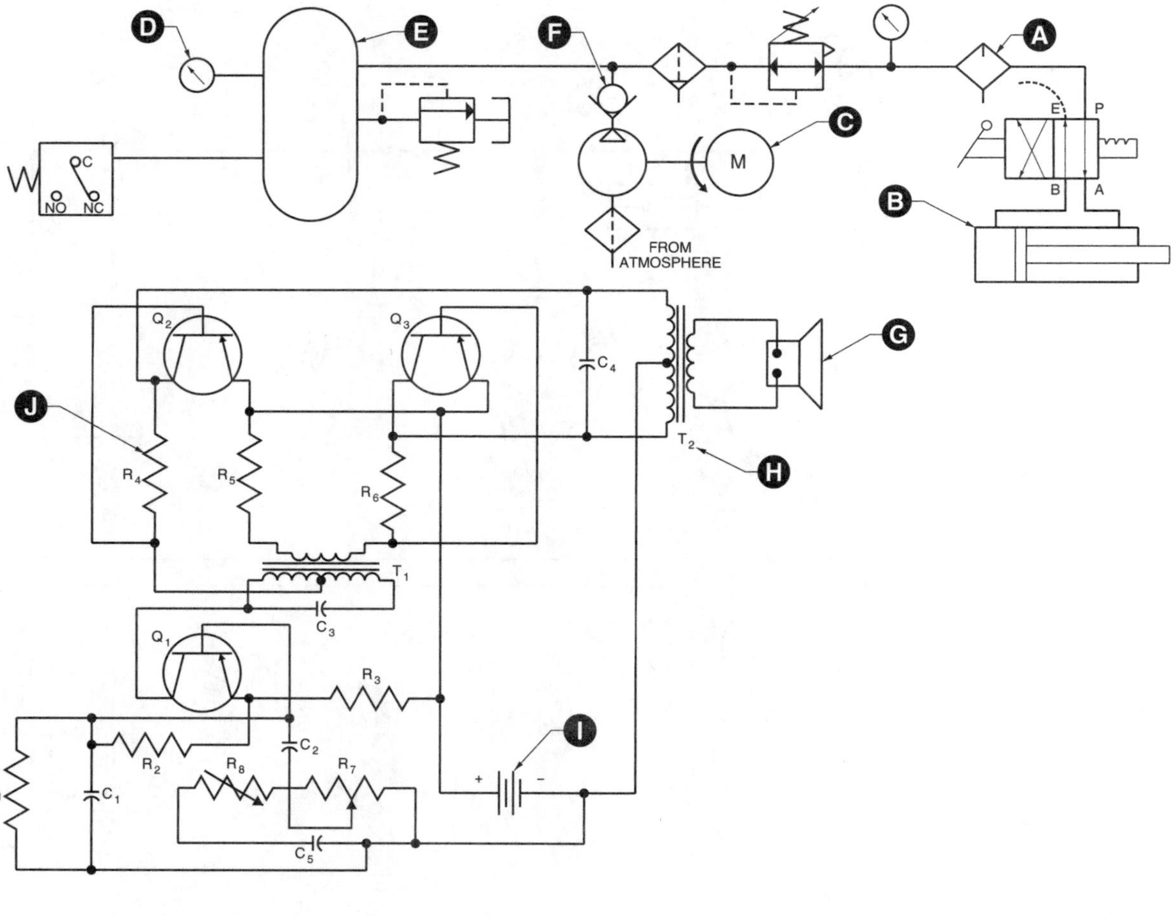

Name _____ Date _____

Hands-On Activity Objectives

- Draw the symbol for each component included in the component kit.
- State the major function of each component included in the component kit.

Procedure 10-1

1. Draw the symbol for a battery.

2. What is the major function of a battery?

3. Draw the symbol for a resistor.

4. What is the major function of a resistor?

5. Draw the symbol for a potentiometer.

6. What is the major function of a potentiometer?

7. Draw the symbol for a DC motor.

8. What is the major function of a DC motor?

9. Draw the symbol for a control relay coil.

10. What is the major function of a control relay coil?

11. Draw the symbol for a control relay normally closed contact.

12. What is the major function of a control relay normally closed contact?

13. Draw the symbol for a control relay normally open contact.

14. What is the major function of a control relay normally open contact?

15. Draw the symbol for a buzzer.

16. What is the major function of a buzzer?

17. Draw the symbol for a normally open pushbutton.

18. What is the major function of a normally open pushbutton?

19. Draw the symbol for normally open limit switch contacts.

20. What is the major function of normally open limit switch contacts?

21. Draw the symbol for normally closed limit switch contacts.

22. What is the major function of normally closed limit switch contacts?

23. Draw the symbol for a red lamp.

24. What is the major function of a red lamp?

25. Draw the symbol for a yellow lamp.

26. What is the major function of a yellow lamp?

27. Draw the symbol for a green lamp.

28. What is the major function of a green lamp?

29. Draw the symbol for a diode.

30. What is the major function of a diode?

Printreading Objectives

• Read an industrial electrical print and identify the symbols being used on the print.

Procedure 10-2

Click on the Compactor Print button on the CD-ROM and answer the following questions. Line 19 includes the timer coil for timer one (1TR). The 1TR contacts are located in line 10.

_____ **1.** Is 1TR an ON-delay or OFF-delay timer?

_____ **2.** What color is the "container full" light in line 25?

_____ **3.** What type of electrical switch is shown in line 25?

_____ **4.** Is the electrical switch in line 25 normally open or normally closed?

_____ **5.** The oil heater in line 1 is controlled by what type of switch?

_____ **6.** Is the switch that controls the oil heater normally open or normally closed?

_____ **7.** Is the compactor motor a DC, single-phase, or three-phase motor?

Measurement Activity Objectives

- Use a DMM to check the operation of a switch.
- Use a DMM to take voltage and current measurement in electrical circuits.

Procedure 10-3

Build the following circuit and take measurements. Set the DMM to measure VDC.

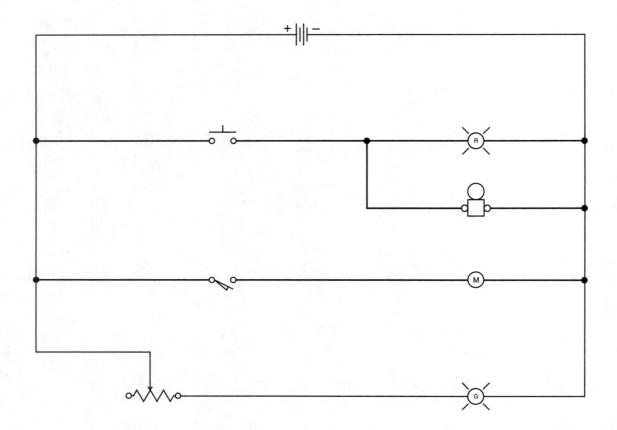

_____ **1.** What is the voltage of the power supply?

_____ **2.** What happens to the voltage across the green lamp when the potentiometer is rotated?

Set the DMM to measure in-line mA DC.

_____ **3.** What is the current of the motor?

_____ **4.** What happens to the current of the green lamp as the potentiometer dims the lamp?

Name _____ **Date** _____

True-False

T	F	**1.**	Insulated conductors are rated based on the current they may safely carry without damaging the insulation.
T	F	**2.**	A cold solder joint is a good electrical and mechanical connection caused by sufficient heat during soldering.
T	F	**3.**	Wire is sized by using a number, such as No. 12 or No. 14.
T	F	**4.**	An overload is a condition that exists on an electrical circuit when the normal load current is exceeded.
T	F	**5.**	A circuit breaker is an OCPD with a mechanical mechanism that may manually or automatically open the circuit when an overload condition or short circuit occurs.
T	F	**6.**	A ripcord is a cord included in a cable that aids in removing the outer jacket.

Multiple Choice

_____ **1.** ___ is any individual conductor.
  A. Wire
  B. Nonmetallic cable
  C. Patch cord
  D. Fiber-optic cable

_____ **2.** ___ is two or more conductors grouped together within a common protective cover and used to connect individual components.
  A. Wire
  B. Cable
  C. Conduit
  D. Patch cord

_____ **3.** ___ is two or more conductors grouped together and used to deliver power to a load by means of a plug.
    A. Wire
    B. Cable
    C. Cord
    D. Conduit

_____ **4.** ___ is the method of using light to transport information from one location to another through thin filaments of glass or plastic.
    A. Cladding
    B. Soldering
    C. A pigtail splice
    D. Fiber optics

_____ **5.** A ___ splice is a splice that connects two wires together when one wire is to remain unbroken.
    A. pigtail
    B. Western Union
    C. standard tap
    D. knotted tap

Completion

_____ **1.** A(n) ___ is a material that has very little resistance and permits electrons to move through it easily.

_____ **2.** ___ is a hollow pipe used to protect conductors.

_____ **3.** ___ cable is a tough, plastic-sheathed cable that is normally made of a moisture-resistant, flame-retardant material.

_____ **4.** A(n) ___ is a tool that is designed to properly remove insulation from small gauge (normally AWG sizes No. 10 – No. 22) wires.

_____ **5.** ___ cable is a cable that has an outer cover made of flexible galvanized steel.

_____ **6.** A(n) ___ is a solderless plastic connector that uses a tapered metal coil spring to twist wires together.

_____ **7.** ___ is the process of joining metals by heat to make a strong electrical and mechanical connection.

_____ **8.** A(n) ___ splice is a splice that consists of twisting two wires together.

_____ **9.** A(n) ___ splice is a splice that connects two wires that may be placed under a lot of mechanical strain.

_____ **10.** A(n) ___ is the device at the end of a cord that connects the device to the electrical power supply by means of a receptacle.

_____ **11.** A(n) ___ plug is a plug in which one blade is wider than the other blade.

_____ **12.** A(n) ___ is overcurrent that leaves the normal current-carrying path by going around the load and back to the power source or ground.

_____ **13.** A(n) ___ connection is a connection in which two wires are held together by crimping them in a specially-designed fitting.

_____ **14.** A(n) ___ is an OCPD with a fusible link that melts and opens the circuit when an overload condition or short circuit occurs.

_____ **15.** A(n) ___ is the condition that occurs when circuit current rises above the normal current level at which the load and/or circuit is designed to operate.

_____ **16.** A(n) ___ fuse is a fuse that may detect an overcurrent and open the circuit almost instantly.

_____ **17.** A(n) ___ fuse is a fuse that may detect and remove a short circuit almost instantly, but allow small overloads to exist for a short period of time.

_____ **18.** A(n) ___ bend consists of two 45° angles on one piece of conduit.

_____ **19.** ___ cable is used to transmit data from one location within a system to another location.

_____ **20.** ___ is the range of frequencies that a device can accept within tolerable limits.

Wire Splices

_____ **1.** Western Union

_____ **2.** Pigtail

_____ **3.** Fixture

_____ **4.** Tap

Circuit Conductors, Connections, and Protection 11

Worksheet 11-1

Name ______________________________ **Date** ________________

Conductor and Conduit Identification

________________ **1.** Ribbon cable

________________ **2.** Wire

________________ **3.** Flexible conduit

________________ **4.** Armored cable (BX)

________________ **5.** Coaxial cable

________________ **6.** EMT conduit

________________ **7.** Nonmetallic-sheathed cable (Romex)

________________ **8.** Steel conduit

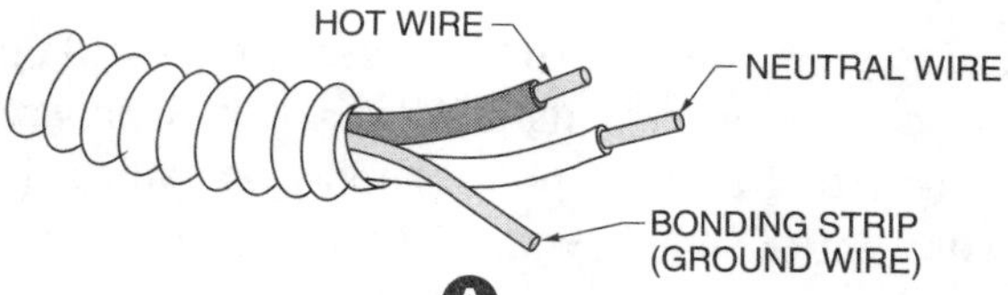
HOT WIRE
NEUTRAL WIRE
BONDING STRIP
(GROUND WIRE)

A

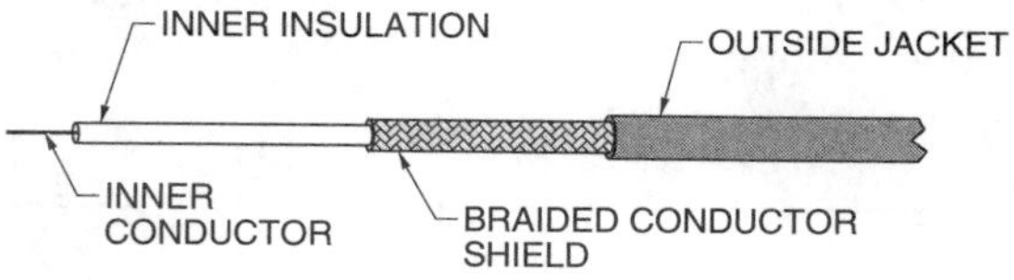
INNER INSULATION
OUTSIDE JACKET
INNER
CONDUCTOR
BRAIDED CONDUCTOR
SHIELD

B

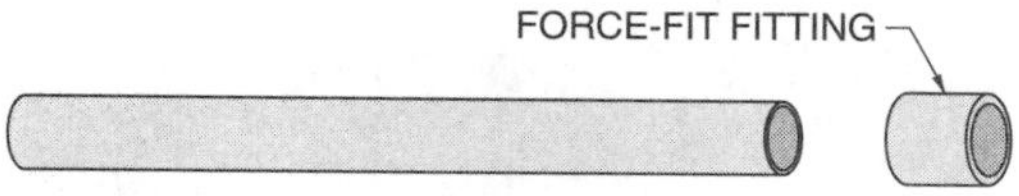
FORCE-FIT FITTING

C

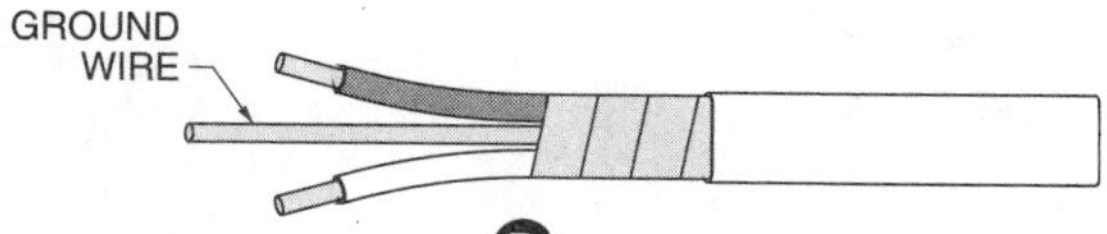
GROUND
WIRE

D

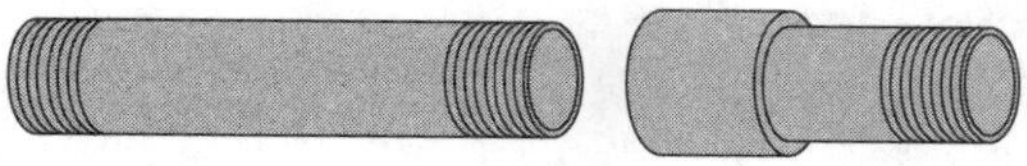

E

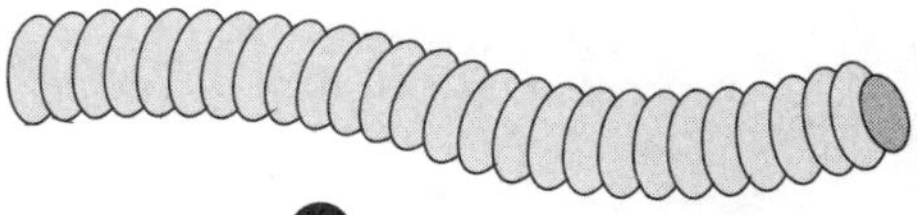

F

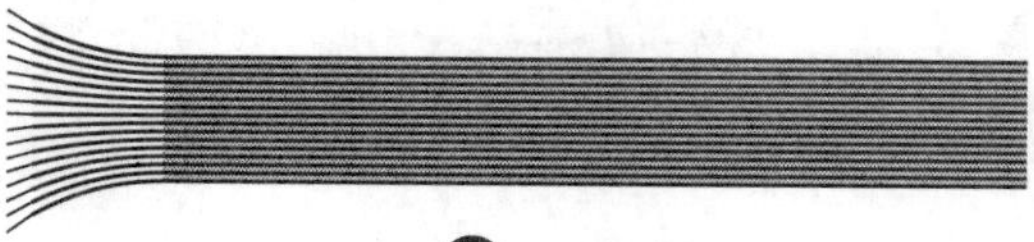

G

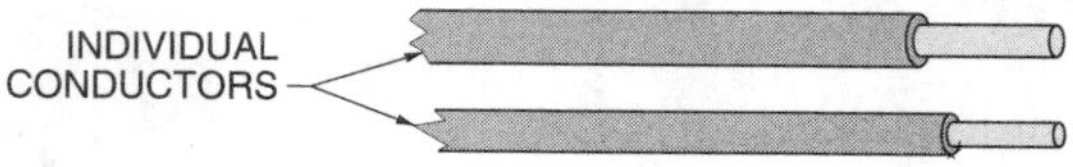
INDIVIDUAL
CONDUCTORS

H

Name _____ **Date** _____

Fuses and Circuit Breakers

1. Every ungrounded (hot) power line must be protected against short circuits and overloads. Connect loads 1 and 2 to the proper power supply by adding the correct number of switch(es) and fuse(s) inside each enclosure and connecting these to the correct power supply. Connect load 3 to the proper power supply by adding the correct number of switch(es) and circuit breaker(s) inside the enclosure and connecting these to the correct power supply.

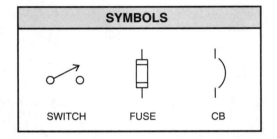

SYMBOLS		
SWITCH	FUSE	CB

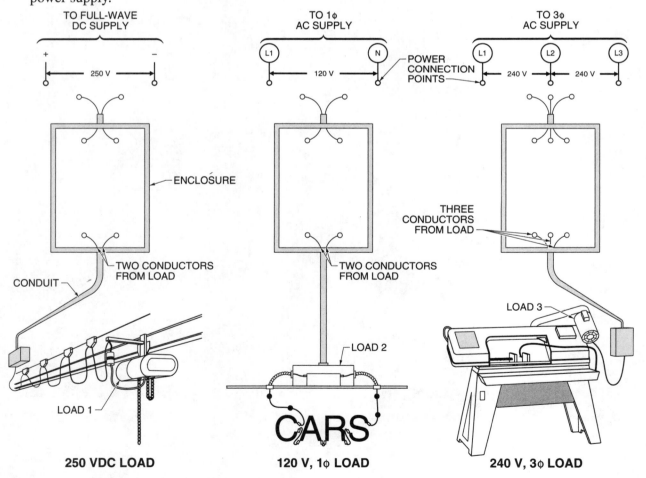

TO FULL-WAVE DC SUPPLY

+   250 V   −

ENCLOSURE

CONDUIT

TWO CONDUCTORS FROM LOAD

LOAD 1

250 VDC LOAD

TO 1φ AC SUPPLY

L1   120 V   N

POWER CONNECTION POINTS

TWO CONDUCTORS FROM LOAD

LOAD 2

CARS

120 V, 1φ LOAD

TO 3φ AC SUPPLY

L1   240 V   L2   240 V   L3

THREE CONDUCTORS FROM LOAD

LOAD 3

240 V, 3φ LOAD

Circuit Conductors, Connections, and Protection Activities 11

Name _____ Date _____

Printreading Activity Objectives

- Use a print to determine the type of fuse that should be used at a given location in an industrial system.
- Use a print to determine the current path through different fuses in a circuit.

Procedure 11-1

Click on the Compactor Print button on the CD-ROM and answer the following questions.

1. Should NTDF- or TDF-rated fuses be used in the main fused disconnect switch?

2. On the control-circuit side of the transformer (X1 and X2), why is there no fuse on the X2 side?

3. On the power-circuit side of the transformer (H1 and H4), why are both the H1 and H4 conductors fused?

4. What determines the current rating of fuse 2FU?

5. Does the current of the oil heater flow through fuse 1FU?

Wire Sizing Activity Objectives

- Determine the amount of current drawn by lamps over the basketball court area.
- Determine the total amount of current drawn by all lamps in the gymnasium.
- Determine the correct wire size for the main feeder conductors into the lighting panel.

Procedure 11-2

Metal-halide lamps are used to light a gymnasium. A total of 54 lamps are used in the gymnasium. Twenty-eight 150 W lamps are used over the basketball court, and 26 100 W lamps are used around the perimeter of the court. Click on the Lamp Data Sheets button on the CD-ROM. Next, click on the Metal Halide Lamp button.

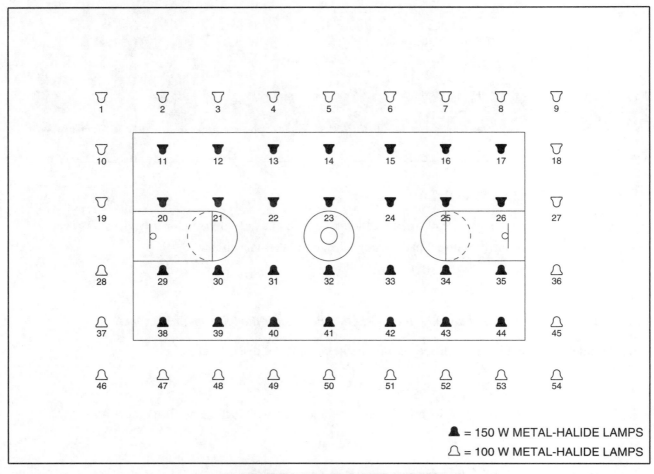

BASKETBALL COURT LIGHTING

▲ = 150 W METAL-HALIDE LAMPS
△ = 100 W METAL-HALIDE LAMPS

_____ **1.** What is the specified current of a 100 W lamp?

_____ **2.** What is the specified current of a 150 W lamp?

_____ **3.** What is the total current draw of all the lamps in the gymnasium?

Click on the Ampacities of Insulated Conductors button on the CD-ROM.

_____ **4.** Assume each conductor must be rated to carry the full current of all lamps. Without derating for ambient temperature or number of conductors in a conduit, an AWG # ___ size THHW copper wire is required for the main feeder conductors into the lighting panel that subdivides the lamps into smaller branch circuits.

_____ **5.** If a correction factor for a 38°C (100.4°F) ambient temperature must be applied, an AWG # ___ size THHW copper wire should be used.

Series Circuits

Review Questions

12

Name _____ Date _____

True-False

T   F   **1.** A series connection is a connection that has two or more components connected so there is only one path for current flow.

T   F   **2.** The voltage applied across loads connected in series is divided across the loads.

T   F   **3.** In a DC series circuit, each component (switch, load, fuse, etc.) has a positive polarity side and a negative polarity side.

T   F   **4.** The lower the power rating of a load, the lower the resistance value.

T   F   **5.** The current in a circuit containing series-connected loads is the same throughout the circuit.

T   F   **6.** In a series circuit, the higher the load's resistance or applied voltage, the less power produced.

T   F   **7.** A capacitor consists of two metal plates (conductors) separated by a dielectric (insulating material).

T   F   **8.** A photocell is a voltage source that converts light energy to electrical energy.

Multiple Choice

_____ **1.** The unit used to measure capacitance is the ___.
A. ohm (Ω)
B. farad (F)
C. henry (H)
D. volt (V)

_____ **2.** ___ is the ability to store energy in the form of an electrical charge.
A. Resistance
B. Capacitance
C. Inductance
D. Impedance

_____ **3.** Small conductors have a higher ___ than large conductors made of the same material.
    A. voltage
    B. resistance
    C. capacitance
    D. inductance

_____ **4.** In an ideal circuit, the loads should be the only devices in the circuit that have ___.
    A. resistance
    B. voltage
    C. capacitance
    D. current

_____ **5.** Switches are usually connected in series to build ___ in a circuit.
    A. safety
    B. pressure
    C. resistance
    D. speed

Completion

_____ **1.** ___ is the positive (+) or negative (–) state of an object.

_____ **2.** ___ is the opposition to current flow.

_____ **3.** Loads are normally represented in a schematic diagram by the ___ symbol.

_____ **4.** A(n) ___ is an electric device designed to store electrical energy by means of an electrostatic field.

_____ **5.** ___ is the property of an electric device that opposes a change in current due to its ability to store electrical energy stored in a magnetic field.

_____ **6.** A(n) ___ is a DC voltage source that converts chemical energy to electrical energy.

_____ **7.** A(n) ___core inductor consists of a coil of wire wrapped around a hollow core.

_____ **8.** A(n) ___core inductor consists of a coil of wire wrapped around a laminated steel core.

_____ **9.** A(n) ___core inductor consists of a coil of wire wrapped around a ceramic material.

_____ **10.** A(n) ___ is a unit that produces electricity at a fixed voltage and current level.

Series Circuit Calculations

_____ **1.** $R_T =$ ___ Ω

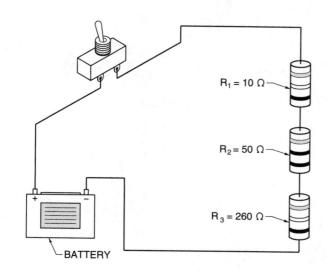

_____ **2.** $E_T =$ ___ V

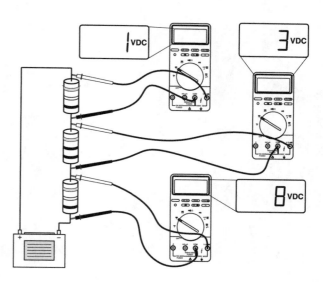

_____ **3.** $P_T =$ ___ W

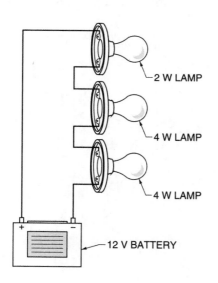

_____ **4.** $I_T =$ ___ A

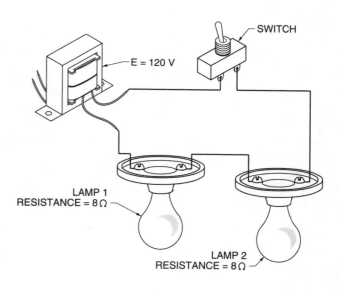

Name _____ Date _____

Resistance in Series Circuits

_____ 1. R_T = ___ Ω

_____ 2. R_T = ___ kΩ

_____ 3. R_T = ___ MΩ

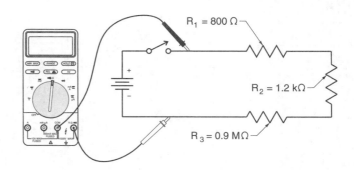

_____ 4. R_T = ___ Ω

_____ 5. R_T = ___ kΩ

_____ 6. R_T = ___ MΩ

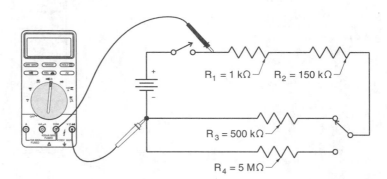

_____ 7. R_T = ___ Ω

_____ 8. R_T = ___ kΩ

_____ 9. R_T = ___ MΩ

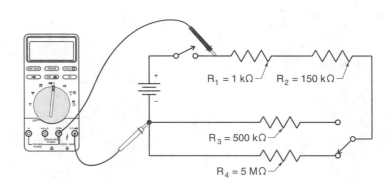

Name _____ **Date** _____

Voltage in Series Circuits

_____ **1.** E = ___ V

_____ **2.** E = ___ mV

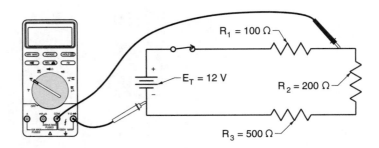

_____ **3.** E = ___ V

_____ **4.** E = ___ kV

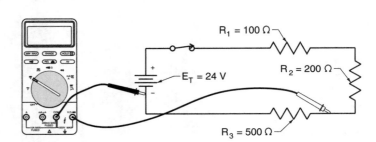

_____ **5.** E = ___ V

_____ **6.** E = ___ kV

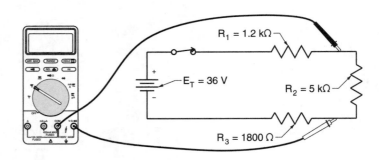

_____ **7.** E = ___ V

_____ **8.** E = ___ mV

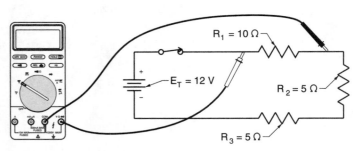

Name _____ **Date** _____

Current in Series Circuits

_____ 1. $I_T =$ ___ A

_____ 2. $I_T =$ ___ mA

_____ 3. $I_T =$ ___ μA

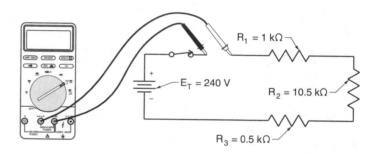

_____ 4. $I_T =$ ___ A

_____ 5. $I_T =$ ___ mA

_____ 6. $I_T =$ ___ μA

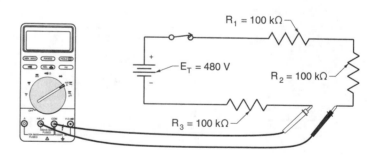

_____ 7. $I_T =$ ___ A

_____ 8. $I_T =$ ___ mA

_____ 9. $I_T =$ ___ μA

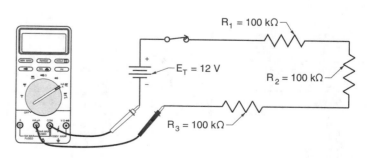

_____ 10. $I_T =$ ___ A

_____ 11. $I_T =$ ___ mA

_____ 12. $I_T =$ ___ μA

Name _____ **Date** _____

Power in Series Circuits

_____ **1.** $P_T =$ ___ W

_____ **2.** $P_T =$ ___ kW

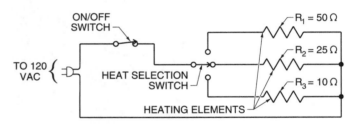

_____ **3.** $P_T =$ ___ W

_____ **4.** $P_T =$ ___ kW

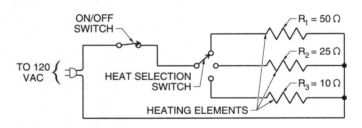

_____ **5.** $P_T =$ ___ W

_____ **6.** $P_T =$ ___ kW

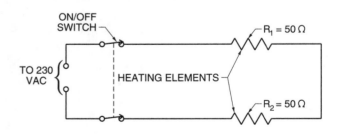

_____ **7.** $P_T =$ ___ W

_____ **8.** $P_T =$ ___ kW

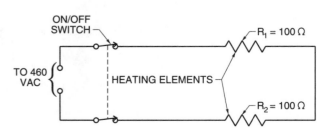

_____ **9.** $P_T =$ ___ W

_____ **10.** $P_T =$ ___ kW

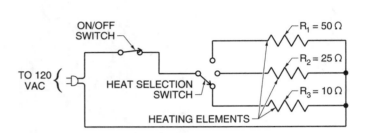

Name _____ **Date** _____

Capacitors and Inductors in Series Circuits

_____ **1.** C_T = ___ F

_____ **2.** C_T = ___ μF

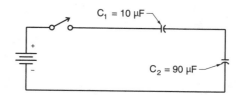

_____ **3.** C_T = ___ F

_____ **4.** C_T = ___ μF

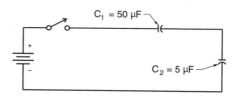

_____ **5.** C_T = ___ F

_____ **6.** C_T = ___ μF

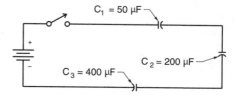

_____ **7.** L_T = ___ H

_____ **8.** L_T = ___ μH

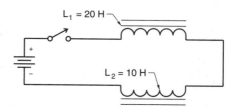

_____ **9.** L_T = ___ H

_____ **10.** L_T = ___ kH

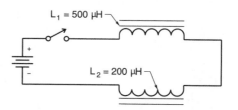

Name _____ Date _____

Batteries and Solar Cells in Series Circuits

_____ 1. E_T = ___ V

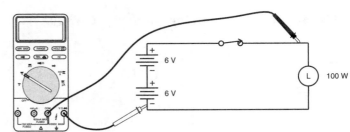

_____ 2. E_T = ___ V

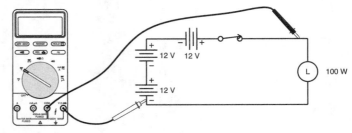

_____ 3. E = ___ V

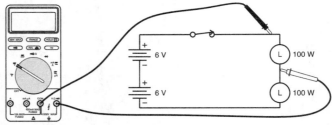

_____ 4. E = ___ V

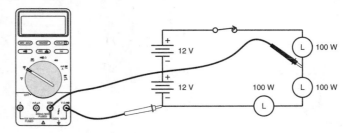

_____ 5. E_T = ___ V

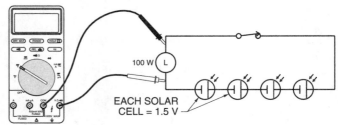

Name _____ Date _____

Drawing Activity Objectives

- Draw a series circuit using standard electrical symbols. See Industrial Electrical Symbols in the Appendix.
- Calculate the total resistance of a series circuit.
- Calculate the total current of a series circuit.
- Calculate the voltage across each component of a series circuit.

Procedure 12-1

1. Draw a circuit using standard electrical symbols that includes an ON/OFF toggle switch, a 100 Ω resistor, and a 150 Ω resistor connected in series to a 9 VDC power supply. Label the 100 Ω resistor R_1 and the 150 Ω resistor R_2.

_____ 2. The total resistance of the series circuit is ___ Ω.

_____ 3. The total current of the series circuit is ___ mA.

_____ 4. The voltage across R_1 is ___ VDC.

_____ 5. The voltage across R_2 is ___ VDC.

_____ 6. If the 100 Ω resistor is replaced with a 300 Ω resistor, will the voltage across the 150 Ω resistor increase or decrease? *Note:* Calculate the new total current and voltage drop across each resistor to determine the answer.

Hands-On Activity Objectives

- Connect the components of a series circuit together and observe how the brightness of a lamp is affected by higher and lower resistances connected in series with the lamp.

Procedure 12-2

Using the Electrical Component Identification card in the Components Kit, identify the battery, ON/OFF toggle switch, yellow lamp, 150 Ω resistor, and test leads. Build the circuit of Procedure 12-1 substituting the yellow lamp for R_1. The yellow lamp should turn ON but will be dim when the switch is in the ON position.

1. Why is the yellow lamp dim and not at full brightness?

2. What happens to the brightness of the lamp if the 150 Ω resistor R_2 is replaced with a 75 Ω resistor? Why?

3. What happens to the brightness of the lamp if resistor R_2 is replaced with a 200 Ω resistor? Why?

Measurement Activity Objectives

- Calculate the amount of voltage, current, and resistance at different points in a series circuit.
- Measure the total resistance of a series circuit.
- Measure the voltage across each component in a series circuit.
- Measure the current in a series circuit.

Procedure 12-3

Build the circuit of Procedure 12-1 using a 100 Ω resistor for R_1 and a 1000 Ω resistor for R_2. With the battery disconnected, measure the total resistance.

_____ **1.** Total measured resistance is ___ Ω.

With power ON and the DMM set to measure VDC, measure the voltage across R_1 and R_2.

_____ **2.** Voltage measured across R_1 is ___ VDC.

_____ **3.** Voltage measured across R_2 is ___ VDC.

With the DMM set to measure in-line current, measure the total current.

_____ **4.** Total measured current is ___ mA.

Parallel Circuits

Review Questions

13

Name _____ **Date** _____

True-False

T   F   **1.** Care must be taken when working with parallel circuits because current can be flowing in one part of the circuit even though another part of the circuit is turned OFF.

T   F   **2.** All DC voltage sources have a positive and a negative terminal.

T   F   **3.** All parallel-connected switches must be closed to start current flow.

T   F   **4.** The total resistance in a circuit containing parallel-connected loads is more than the smallest resistance value.

T   F   **5.** All parallel-connected switches must be opened to stop current flow.

T   F   **6.** The total resistance decreases if loads are added in parallel and increases if loads are removed.

T   F   **7.** When one branch of a parallel circuit is short circuited, all other paths are also short circuited.

T   F   **8.** The current capacity of a battery is decreased by connecting cells in parallel.

T   F   **9.** Total current in a circuit containing parallel-connected loads equals the sum of the current through all the loads.

T   F   **10.** Total current decreases if loads are added in parallel and increases if loads are removed.

Multiple Choice

_____ **1.** ___ is the property of an electric device that opposes a change in current due to its ability to store electrical energy in a magnetic field.
   A. Capacitance
   B. Power
   C. Inductance
   D. Resistance

_____ 2. Power produced by a load is equal to the voltage drop across the load ___ the current through the load.
      A. divided by
      B. times
      C. plus
      D. minus

_____ 3. Total ___ is equal to the sum of the power produced by each load.
      A. voltage
      B. power
      C. current
      D. resistance

_____ 4. A loose connection increases the circuit resistance and decreases the circuit ___.
      A. voltage
      B. power
      C. current
      D. inductance

_____ 5. A(n) ___ is a voltage source that converts light energy into electrical energy.
      A. battery
      B. solar cell
      C. inductor
      D. capacitor

Completion

_____ 1. A(n) ___ connection is a connection that has two or more components connected so that there is more than one path for current flow.

_____ 2. Connecting loads in ___ is the most common method used to connect loads.

_____ 3. When designing a circuit, the size of wire used is based on the amount of expected ___.

_____ 4. The equivalent capacitance of capacitors connected in ___ is equal to the sum of the individual capacitors.

_____ 5. The voltage across each load is the same when loads are connected in ___.

Parallel Circuit Calculations

_____ **1.** $R_T =$ ___ Ω

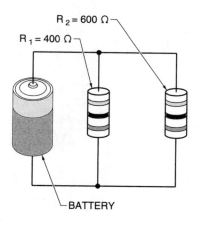

$R_2 = 600\ \Omega$
$R_1 = 400\ \Omega$

BATTERY

_____ **2.** $E_T =$ ___ V

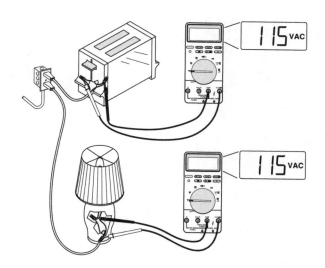

| 115 | VAC |

| 115 | VAC |

_____ **3.** $I_T =$ ___ A

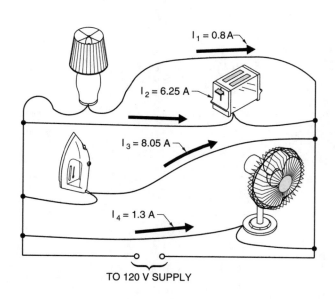

$I_1 = 0.8\ A$

$I_2 = 6.25\ A$

$I_3 = 8.05\ A$

$I_4 = 1.3\ A$

TO 120 V SUPPLY

_____ **4.** $P_T =$ ___ W

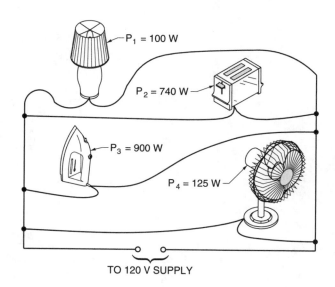

$P_1 = 100\ W$

$P_2 = 740\ W$

$P_3 = 900\ W$

$P_4 = 125\ W$

TO 120 V SUPPLY

Parallel Circuits

Worksheet 13-1

13

Name _____ Date _____

Resistance in Parallel Circuits

_____ **1.** $R_T =$ ___ Ω

_____ **2.** $R_T =$ ___ $k\Omega$

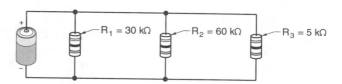

_____ **3.** $R_T =$ ___ Ω

_____ **4.** $R_T =$ ___ $k\Omega$

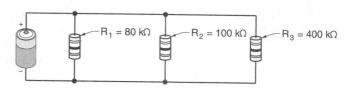

_____ **5.** $R_T =$ ___ Ω

_____ **6.** $R_T =$ ___ $k\Omega$

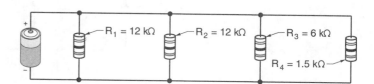

_____ **7.** $R_T =$ ___ Ω

_____ **8.** $R_T =$ ___ $k\Omega$

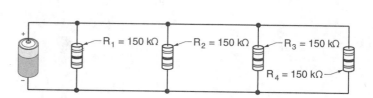

_____ **9.** $R_T =$ ___ Ω

_____ **10.** $R_T =$ ___ $k\Omega$

129

Name _____ Date _____

Current in Parallel Circuits

_____ **1.** Current at A = ___ mA

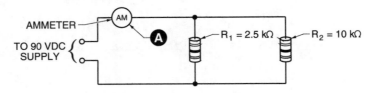

AMMETER — AM Ⓐ
TO 90 VDC SUPPLY
$R_1 = 2.5\ k\Omega$ $R_2 = 10\ k\Omega$

_____ **2.** Current at B = ___ mA

_____ **3.** Current at C = ___ mA

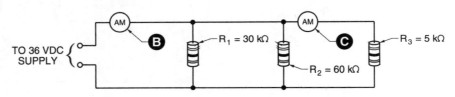

AM Ⓑ   TO 36 VDC SUPPLY $R_1 = 30\ k\Omega$ AM Ⓒ $R_3 = 5\ k\Omega$ $R_2 = 60\ k\Omega$

_____ **4.** Current at D = ___ mA

_____ **5.** Current at E = ___ μA

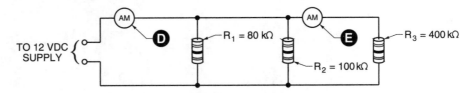

AM Ⓓ   TO 12 VDC SUPPLY $R_1 = 80\ k\Omega$ AM Ⓔ $R_3 = 400\ k\Omega$ $R_2 = 100\ k\Omega$

_____ **6.** Current at F = ___ mA

_____ **7.** Current at G = ___ mA

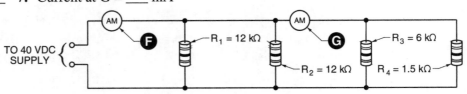

AM Ⓕ   TO 40 VDC SUPPLY $R_1 = 12\ k\Omega$ AM Ⓖ $R_3 = 6\ k\Omega$ $R_2 = 12\ k\Omega$ $R_4 = 1.5\ k\Omega$

Parallel Circuits

Worksheet 13-3

13

Name _____ Date _____

Voltage in Parallel Circuits

_____ **1.** $E_T =$ ___ V

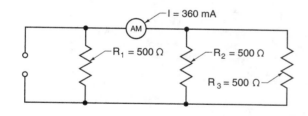

_____ **2.** $E_T =$ ___ V

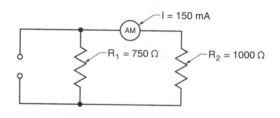

_____ **3.** $E_T =$ ___ V

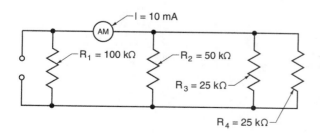

_____ **4.** $E_T =$ ___ V

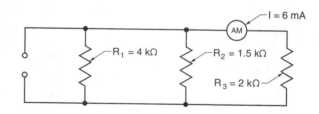

_____ **5.** $E_T =$ ___ V

_____ **6.** $E_T =$ ___ V

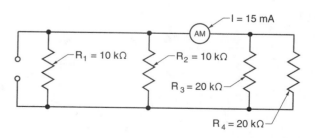

Parallel Circuits

Worksheet 13-4

Name _____ Date _____

Power in Parallel Circuits

_____ 1. $P_T =$ ___ W

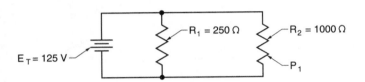

_____ 2. $P_1 =$ ___ W

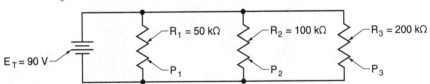

_____ 3. $P_1 =$ ___ mW

_____ 4. $P_2 =$ ___ mW

_____ 5. $P_3 =$ ___ mW

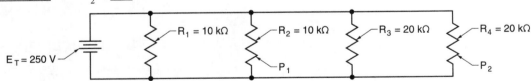

_____ 6. $P_1 =$ ___ W

_____ 7. $P_2 =$ ___ W

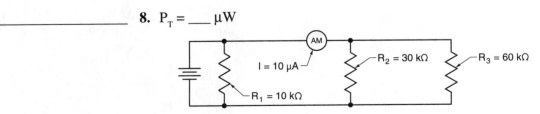

_____ 8. $P_T =$ ___ μW

Name _____ Date _____

Capacitors and Inductors in Parallel Circuits

_____ **1.** $C_T =$ ___ µF

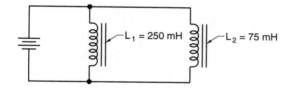

_____ **2.** $C_T =$ ___ µF

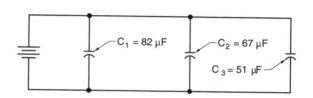

_____ **3.** $L_T =$ ___ mH

_____ **4.** $L_T =$ ___ mH

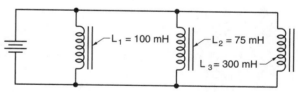

_____ **5.** $L_T =$ ___ mH

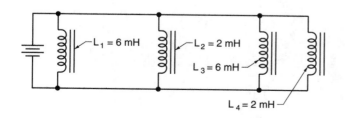

_____ **6.** $L_T =$ ___ mH

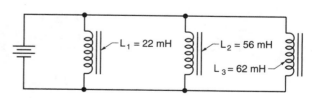

Name _____ **Date** _____

Batteries and Solar Cells in Parallel Circuits

_____ **1.** $R_T =$ ___ Ω

_____ **2.** $E_T =$ ___ V

_____ **3.** $P_T =$ ___ mW

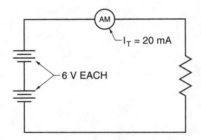

$I_T = 20$ mA

6 V EACH

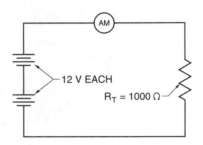

12 V EACH

$R_T = 1000\ \Omega$

_____ **4.** $I_T =$ ___ A

_____ **5.** $I_T =$ ___ mA

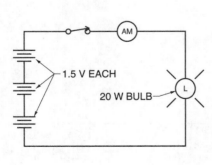

1.5 V EACH

20 W BULB

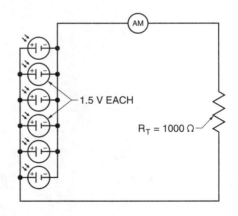

1.5 V EACH

$R_T = 1000\ \Omega$

Name _____ Date _____

Drawing Activity Objectives

- Draw a parallel circuit using standard electrical symbols. See Industrial Electrical Symbols in the Appendix.
- Calculate the total resistance of a parallel circuit.
- Calculate the total current of a parallel circuit.
- Calculate the current through each component in a parallel circuit.

Procedure 13-1

1. Draw a circuit using standard electrical symbols that includes a 1000 Ω resistor, a 10,000 Ω resistor, and a 20,000 Ω resistor connected in parallel to a 9 VDC power supply. Label the 1000 Ω resistor R_1, the 10,000 Ω resistor R_2, and the 20,000 Ω resistor R_3.

_____ 2. The total resistance of the parallel circuit is ___ Ω.

_____ 3. The total current of the parallel circuit is ___ mA.

_____ 4. The current through R_1 is ___ mA.

_____ 5. The current through R_2 is ___ mA.

_____ 6. The current through R_3 is ___ mA.

_____ 7. Which resistor has the least amount of power?

_____ 8. Which resistor has the most amount of power?

_____ 9. If the 20,000 Ω resistor is replaced with a 5600 Ω resistor, will the total current increase or decrease?

Hands-On Activity Objectives

• Connect the components of a parallel circuit together and observe how the brightness of lamps is affected by higher and lower resistances connected in parallel with the lamp.

Procedure 13-2

Using the Electrical Component Identification card in the Components Kit, identify the battery and the red, yellow, and green lamps. Build a test circuit in which the red, yellow, and green lamps are connected in parallel with the 9 VDC power supply. Connect the lamps one at a time and answer the following questions.

1. Did the second lamp dim the brightness of the first lamp when it was connected in parallel? Why or why not?

2. Did the third lamp dim the brightness of the other two lamps when it was connected in parallel? Why or why not?

Connect a 20,000 Ω resistor in parallel with the lamps.

3. Do any of the lamps dim? Why or why not?

Measurement Activity Objectives

• Calculate the amount of voltage, current, resistance, and power at different points in a parallel circuit.
• Measure the total resistance of a parallel circuit.
• Measure the current through each component in a parallel circuit.
• Measure the total current in a parallel circuit.

Procedure 13-3

Build the circuit of Procedure 13-1 using a 1000 Ω resistor for R_1, a 10,000 Ω resistor for R_2, and a 20,000 Ω resistor for R_3. With the battery disconnected, measure the total resistance.

_____ 1. Total measured resistance is ___ Ω.

With power ON and the DMM set to measure in-line current, measure the current through R_1, R_2, and R_3.

_____ 2. Current measured through R_1 is ___ mA.

_____ 3. Current measured through R_2 is ___ mA.

_____ 4. Current measured through R_3 is ___ mA.

Name _____ **Date** _____

True-False

T   F   **1.** The majority of all circuits contain series/parallel-connected components.

T   F   **2.** All points in a DC series/parallel circuit have polarity.

T   F   **3.** Two or more switches must be closed before current flows when the switches are connected in series/parallel.

T   F   **4.** Resistors and loads, such as heating elements, are often connected in a series/parallel combination.

T   F   **5.** Capacitors are normally connected in series or in parallel.

T   F   **6.** Any one or more series-connected switches or all parallel-connected switches must be opened to stop current flow.

T   F   **7.** The total resistance in a circuit containing series/parallel-connected resistors equals the equivalent resistance of the series loads and the sum of the parallel combinations.

T   F   **8.** The higher the resistance of any one resistor or equivalent parallel resistance, the lower the voltage drop.

T   F   **9.** Voltage potential and current capacity are decreased when batteries and solarcells are connected in series/parallel combinations.

T   F   **10.** The total voltage applied across resistors (loads) connected in a series/parallel combination is divided across the individual resistors.

T   F   **11.** The higher the amount of current, the less power produced.

T   F   **12.** The total power in a series/parallel circuit is equal to the sum of the power produced by each load or component.

Multiple Choice

_____ 1. The lower the resistance or higher the amount of current, the more ___ produced.
  A. voltage
  B. power
  C. inductance
  D. capacitance

_____ 2. Current is the same in each ___ part of a series/parallel circuit.
  A. series
  B. parallel
  C. series/parallel
  D. forward-biased

_____ 3. Power is produced when current flows through any load or component that has ___.
  A. power
  B. voltage
  C. inductance
  D. resistance

_____ 4. Switches, loads, meters, and ___ can be connected in a series/parallel connection.
  A. fuses
  B. circuit breakers
  C. other electrical components
  D. all of the above

_____ 5. A load may be any device such as a ___.
  A. heating element
  B. motor
  C. solenoid
  D. all of the above

Completion

_____ 1. A(n) ___ connection is a combination of series- and parallel-connected components.

_____ 2. In most circuits, loads such as lamps, motors, and solenoids are normally connected in ___.

_____ 3. Switches, fuses, and circuit breakers that control and monitor the current through the loads are connected in ___.

_____ 4. Voltmeters are connected in ___ when taking load measurements.

_____ 5. Ammeters are connected in ___ with the load.

Series/Parallel Circuit Calculations

_____ **1.** $R_T =$ ___ Ω

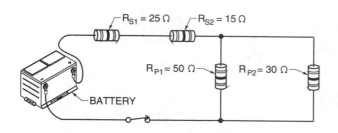

_____ **2.** $E_T =$ ___ V

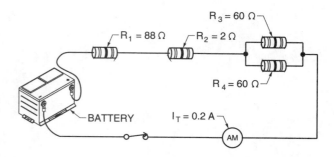

_____ **3.** $I_T =$ ___ A

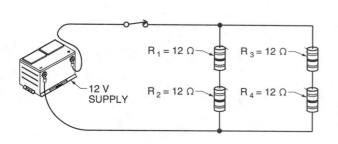

_____ **4.** $P_T =$ ___ W

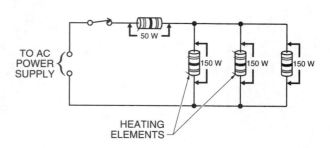

Name _____ **Date** _____

Resistance in Series/Parallel Circuits

_____ **1.** $R_T = $ ___ Ω

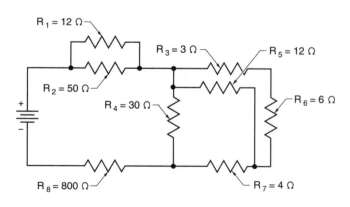

_____ **2.** $R_T = $ ___ Ω

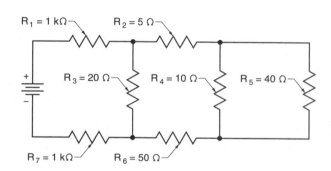

_____ **3.** $R_T = $ ___ Ω

_____ **4.** $R_T = $ ___ Ω

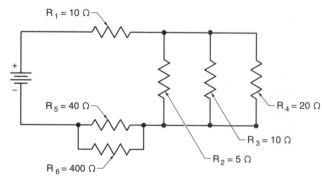

Name _____ **Date** _____

Current in Series/Parallel Circuits

_____ **1.** I_T = ___ mA

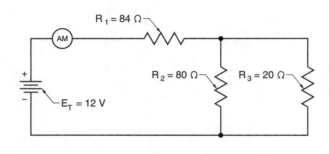

_____ **2.** I_T = ___ mA

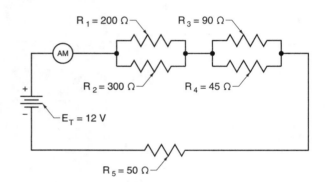

_____ **3.** I_T = ___ mA

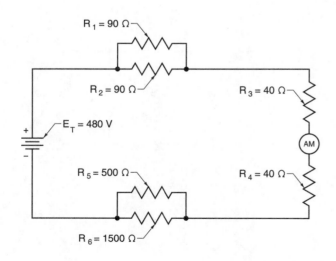

_____ **4.** I_T = ___ mA

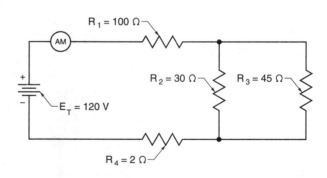

142

Name _____ **Date** _____

Voltage in Series/Parallel Circuits

_____ **1.** E_1 = ___ V   _____ **4.** E_T = ___ V

_____ **2.** E_2 = ___ V

_____ **3.** E_3 = ___ V

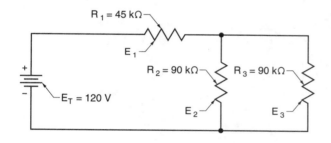

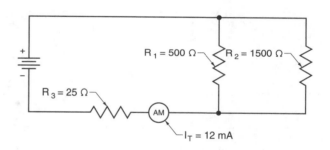

_____ **5.** E_T = ___ V   _____ **7.** E_T = ___ V

_____ **6.** E_4 = ___ V   _____ **8.** E_3 = ___ V

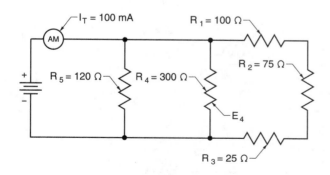

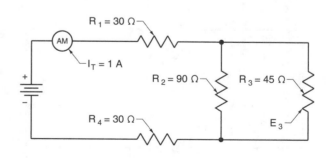

Name _____ **Date** _____

Power in Series/Parallel Circuits

_____ **1.** $P_T =$ ___ W   _____ **3.** $P_T =$ ___ W

_____ **2.** $P_1 =$ ___ W   _____ **4.** $P_2 =$ ___ W

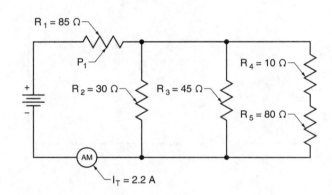

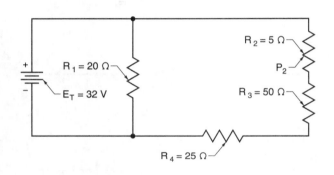

_____ **5.** $P_T =$ ___ W   _____ **7.** $P_T =$ ___ W

_____ **6.** $P_4 =$ ___ W   _____ **8.** $P_4 =$ ___ W

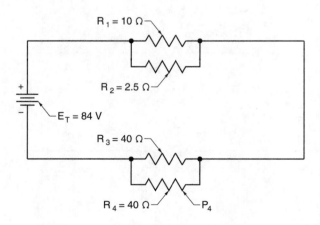

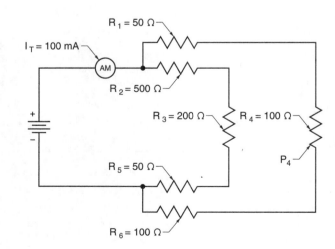

Name _____ Date _____

Power, Resistance, and Current in Series/Parallel Circuits

_____ **1.** $P_T =$ ___ W

_____ **2.** $R_T =$ ___ Ω

_____ **3.** $I_T =$ ___ A

_____ **4.** $I_1 =$ ___ A

_____ **5.** $I_2 =$ ___ A

_____ **6.** $I_3 =$ ___ A

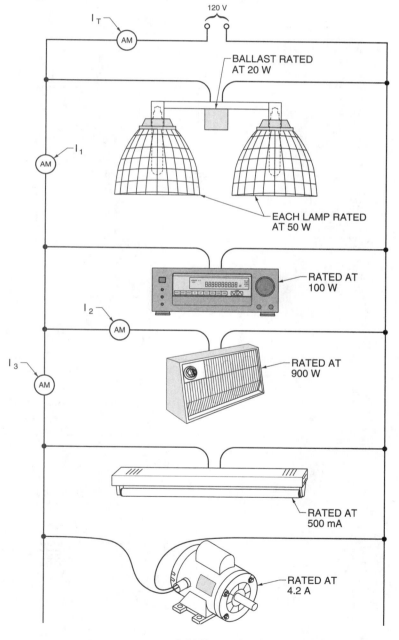

BALLAST RATED AT 20 W

EACH LAMP RATED AT 50 W

RATED AT 100 W

RATED AT 900 W

RATED AT 500 mA

RATED AT 4.2 A

145

Name _____ Date _____

Capacitors and Inductors in Series/Parallel Circuits

_____ **1.** $C_T =$ ___ µF

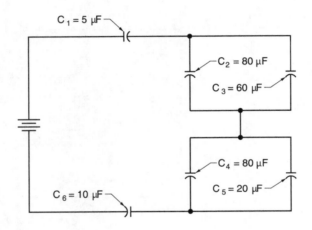

_____ **2.** $L_T =$ ___ H

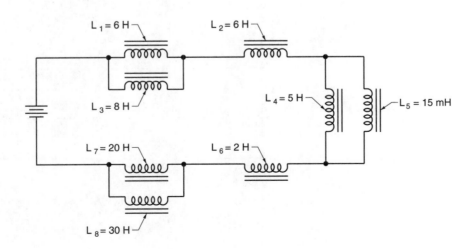

Series/Parallel Circuits

Activities

14

Name _____ Date _____

Drawing Activity Objectives

- Draw a series/parallel circuit using standard electrical symbols. See Industrial Electrical Symbols in the Appendix.
- Calculate the total resistance of a series/parallel circuit.
- Calculate the total current of a series/parallel circuit.
- Calculate the current through each component of a series/parallel circuit.

Procedure 14-1

1. Draw a circuit using standard electrical symbols that includes a 9 VDC power supply, a 20 Ω resistor, a 75 Ω resistor, a 510 Ω resistor, and a 1000 Ω resistor. The 20 Ω resistor (R_1) has one end connected to the positive terminal of a DC power supply and the other end connected in series with one end of the 75 Ω resistor (R_2). The other end of the 75 Ω resistor is connected to the 510 Ω resistor (R_3) and the 1000 Ω resistor (R_4), which are both connected in parallel and have their ends connected to the negative terminal of the power supply.

_____ 2. The total resistance of the series/parallel circuit is ___ Ω.

_____ 3. The total current of the series/parallel circuit is ___ mA.

_____ 4. The power of the series/parallel circuit is ___ mW.

_____ 5. The current through R_1 is ___ mA.

_____ 6. The current through R_3 is ___ mA.

_____ 7. The current through R_4 is ___ mA.

8. If the 20 Ω resistor is replaced with a 100 Ω resistor, will the total current increase or decrease? Why?

Hands-On Activity Objectives

• Connect the components of a series/parallel circuit together and observe how the brightness of the lamps is affected.

Procedure 14-2

Using the Electrical Component Identification card in the Components Kit, identify the battery and the red, yellow, and green lamps. Build a circuit in which the red, yellow, and green lamps are connected in series/parallel with the 9 VDC power supply. Connect the yellow and green lamps in series, and the red lamp in parallel with the yellow and green lamp combination. Connect the lamps to the power supply.

   1. Are all the lamps at full brightness? Why or why not?

   2. Is the current through the yellow lamp the same as the red lamp? Why or why not?

   3. Is the current through the green lamp the same as the current through the yellow lamp? Why or why not?

Measurement Activity Objectives

• Calculate the amount of voltage, current, resistance, and power at different points in a series/parallel circuit.
• Measure the total resistance of a series/parallel circuit.
• Measure the current through each component in a series/parallel circuit.
• Measure the total current in a series/parallel circuit.

Procedure 14-3

Build the circuit of Procedure 14-1 using a 20 Ω resistor for R_1, a 75 Ω resistor for R_2, a 510 Ω resistor for R_3, and a 1000 Ω resistor for R_4. With the battery disconnected, measure the total resistance.

_____ **1.** Total measured resistance is ___ Ω.

With power ON and the DMM set to measure in-line current, measure the current through R_1, R_2, R_3, and R_4.

_____ **2.** The current measured through R_1 is ___ mA.

_____ **3.** The current measured through R_2 is ___ mA.

_____ **4.** The current measured through R_3 is ___ mA.

_____ **5.** The current measured through R_4 is ___ mA.

_____ **6.** The total current measured is ___ mA.

Name _____ **Date** _____

True-False

    T   F   **1.** A solenoid is a device that converts electrical energy into a linear, mechanical force.

    T   F   **2.** Magnetic flux is the invisible lines of force that make up the magnetic field.

    T   F   **3.** The current potential of a transformer is stepped down any time a transformer steps up the voltage.

    T   F   **4.** Hysteresis loss is loss caused by the induced currents that are produced in metal parts that are being magnetized.

    T   F   **5.** Overhead service is electrical service in which service-entrance conductors are run underground from the utility service to the dwelling.

    T   F   **6.** A transformer must be derated if the ambient temperature exceeds 50°C.

Multiple Choice

_____ **1.** ___ is a force that interacts with other magnets and ferromagnetic materials.
        A. Inductance
        B. Impedance
        C. Magnetism
        D. Current

_____ **2.** A ___ is a device that attracts iron and steel because of the molecular alignment of its material.
        A. magnet
        B. transformer
        C. primary coil
        D. power generator

_____ 3. ___ is the magnetic field produced when electricity passes through a conductor.
- A. Electromagnetism
- B. Voltage
- C. Resistance
- D. Current

_____ 4. A ___ is an electric device that uses electromagnetism to change voltage from one level to another or isolate one voltage from another.
- A. generator
- B. smart grid
- C. substation
- D. transformer

_____ 5. ___ is the effect of one coil inducing a voltage into another coil.
- A. Electromagnetism
- B. Impedance
- C. Mutual inductance
- D. Current

_____ 6. The ___ coil of a transformer is the coil to which the voltage is connected.
- A. distribution
- B. primary
- C. secondary
- D. substation

Completion

_____ 1. ___ loss is loss caused by magnetism that remains (lags) in a material after the magnetizing force has been removed.

_____ 2. A(n) ___ transformer is a transformer that dissipates heat through the air surrounding the transformer.

_____ 3. ___ is loss caused by the resistance of copper wire to the flow of current.

_____ 4. ___ is the maximum output required of a transformer.

_____ 5. ___ is the actual power used in an electrical circuit expressed in watts (W).

_____ 6. A(n) ___ configuration is a transformer connection that has one end of each transformer coil connected together.

_____ 7. A(n) ___ is a power delivery system that uses the latest technology to deliver clean, efficient, reliable, and safe power from the utility to its customers.

_____ 8. ___ can only take place with loads that are designed to be remotely controlled.

Name _____ **Date** _____

Transformer Operation

_____ **1.** Primary current = ___ mA

_____ **2.** Primary power = ___ mW

_____ **3.** Secondary voltage = ___ V

_____ **4.** Secondary power = ___ mW

_____ **5.** Transformer voltage ratio =
___ : ___

_____ **6.** Transformer current ratio =
___ : ___

_____ **7.** Transformer power ratio =
___ : ___

_____ **8.** Transformer turns ratio =
___ : ___

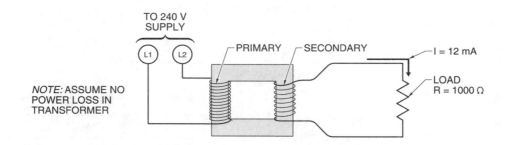

_____ **9.** Primary current = ___ A

_____ **10.** Primary power = ___ W

_____ **11.** Secondary voltage ___ V

_____ **12.** Secondary current = ___ mA

_____ **13.** Secondary power = ___ W

_____ **14.** Transformer current ratio =
___ : ___

_____ **15.** Transformer power ratio =
___ : ___

_____ **16.** Transformer turns ratio =
___ : ___

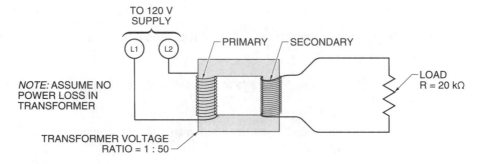

Transformers and Smart Grid Technology

Worksheet 15-2

15

Name _____ Date _____

Transformer Overloading and Temperature Compensation

_____ **1.** A 125 A rated transformer operating on 250 A has a permissible overload time of ___.

_____ **2.** A 72 A rated transformer operating on 216 A has a permissible overload time of ___.

_____ **3.** A 41 A rated transformer operating on 410 A has a permissible overload time of ___.

_____ **4.** A 250 A rated transformer operating on 1000 A has a permissible overload time of ___.

_____ **5.** A 54 A rated transformer operating on 81 A has a permissible overload time of ___.

_____ **6.** A 15 kVA rated transformer installed in an ambient temperature of 45°C has a derated value of ___ kVA.

TRANSFORMER OVERLOADING

TRANSFORMER DERATINGS

Maximum Ambient Temperature (°C)	Maximum Transformer Loading (%)
40	100
45	96
50	92
55	88
60	81
65	80
70	76

_____ **7.** A 30 kVA rated transformer installed in an ambient temperature of 60°C has a derated value of ___ kVA.

_____ **8.** A 50 kVA rated transformer installed in an ambient temperature of 40°C has a derated value of ___ kVA.

_____ **9.** A 100 kVA rated transformer installed in an ambient temperature of 50°C has a derated value of ___ kVA.

_____ **10.** A 125 kVA rated transformer installed in an ambient temperature of 55°C has a derated value of ___ kVA.

Name _____ Date _____

Sizing 1φ Transformers

_____ 1. Required transformer secondary voltage = ___ VAC

_____ 2. Total current rating of all loads = ___ A

_____ 3. Total capacity of transformer = ___ kVA

_____ 4. Total capacity of transformer including safety factor (10%) = ___ kVA

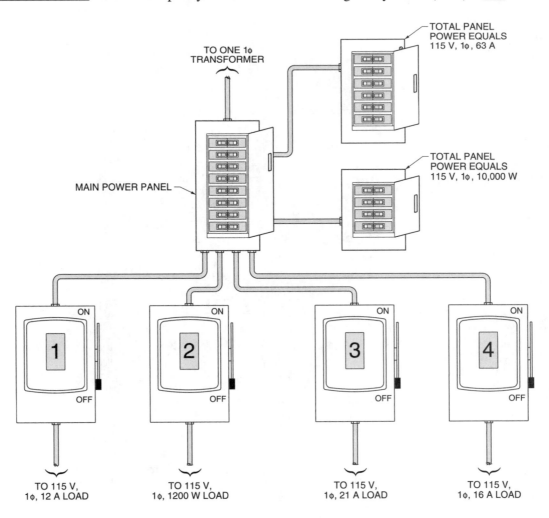

TO ONE 1φ TRANSFORMER

TOTAL PANEL POWER EQUALS 115 V, 1φ, 63 A

TOTAL PANEL POWER EQUALS 115 V, 1φ, 10,000 W

MAIN POWER PANEL

ON 1 OFF

ON 2 OFF

ON 3 OFF

ON 4 OFF

TO 115 V, 1φ, 12 A LOAD

TO 115 V, 1φ, 1200 W LOAD

TO 115 V, 1φ, 21 A LOAD

TO 115 V, 1φ, 16 A LOAD

Name _____ **Date** _____

Sizing 3ϕ Transformers

_____ **1.** Total current rating of all loads = ___ A

_____ **2.** Total capacity of transformer bank = ___ kVA

_____ **3.** Total capacity of transformer including safety factor (10%) = ___ kVA

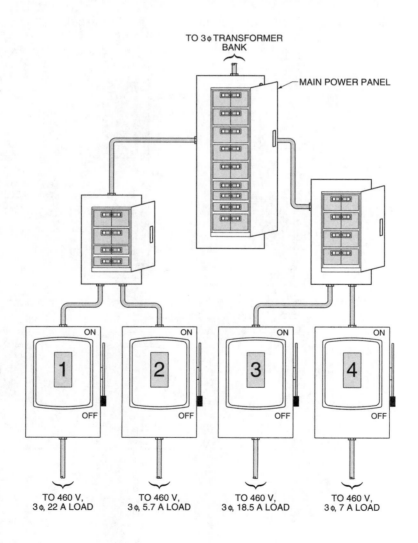

TO 3ϕ TRANSFORMER BANK

— MAIN POWER PANEL

ON **1** OFF	ON **2** OFF	ON **3** OFF	ON **4** OFF

TO 460 V, 3ϕ, 22 A LOAD

TO 460 V, 3ϕ, 5.7 A LOAD

TO 460 V, 3ϕ, 18.5 A LOAD

TO 460 V, 3ϕ, 7 A LOAD

Name _____ **Date** _____

Transformer Current Draw

_____ **1.** A 0.75 kVA, 120 V, 1ϕ transformer loaded at 90% draws ___ A.

_____ **2.** A 15 kVA, 120 V, 1ϕ transformer loaded at 100% draws ___ A.

_____ **3.** A 7.5 kVA, 120 V, 1ϕ transformer loaded at 85% draws ___ A.

_____ **4.** A 5 kVA, 120 V, 1ϕ transformer loaded at 120% draws ___ A.

_____ **5.** A 10 kVA, 120 V, 1ϕ transformer loaded at 50% draws ___ A.

_____ **6.** A 1.5 kVA, 120 V, 1ϕ transformer loaded at 105% draws ___ A.

_____ **7.** A 6 kVA, 230 V, 3ϕ transformer loaded at 75% draws ___ A.

_____ **8.** A 9 kVA, 230 V, 3ϕ transformer loaded at 40% draws ___ A.

_____ **9.** A 15 kVA, 230 V, 3ϕ transformer loaded at 85% draws ___ A.

_____ **10.** A 22 kVA, 230 V, 3ϕ transformer loaded at 125% draws ___ A.

_____ **11.** A 30 kVA, 230 V, 3ϕ transformer loaded at 60% draws ___ A.

_____ **12.** A 45 kVA, 230 V, 3ϕ transformer loaded at 70% draws ___ A.

_____ **13.** A 4 kVA, 230 V, 3ϕ transformer loaded at 105% draws ___ A.

_____ **14.** A 9 kVA, 230 V, 3ϕ transformer loaded at 92% draws ___ A.

120 V, 1ϕ TRANSFORMER

METER SOCKET

230 V, 3ϕ TRANSFORMER

Name _____ Date _____

Transformer Connections

1. Connect the power line terminals to the transformer primary terminals in a delta configuration. Connect the load terminals to the transformer secondary terminals in a wye configuration.

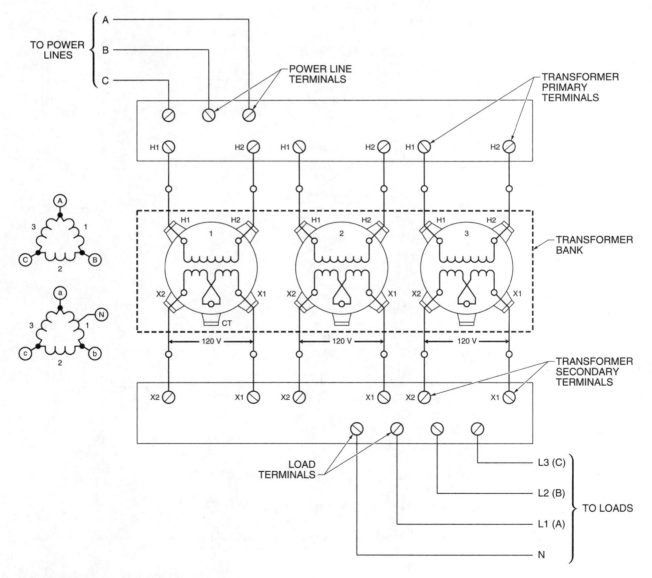

Transformers and Smart Grid Technology

Activities

15

Name _____ Date _____

Printreading Activity Objectives

- Determine the correct wiring of a control transformer used in an electrical machine.
- Determine the required size of a control transformer used in an electrical machine.

Procedure 15-1

Click on the Compactor Print button on the CD-ROM and complete the following statements.

_____ 1. The transformer used in this circuit is a step-___ transformer.

_____ 2. If the applied system voltage at L_1, L_2, and L_3 is 480 VAC, the two transformer terminals that L_1 and L_3 should be connected to after the line fuses are ___ and ___.

_____ 3. With 480 VAC primary voltage, the transformer voltage ratio is ___ to ___.

_____ 4. With 480 VAC primary voltage, the transformer current ratio is ___ to ___.

_____ 5. With 480 VAC primary voltage, the transformer power ratio is ___ to ___.

_____ 6. With 480 VAC primary voltage, the transformer turns ratio is ___ to ___.

_____ 7. If the loads (solenoids, lamps, relay coils, motor starter coil, and timer coils) have a total current rating of 6.4 A, the kVA capacity rating of the transformer after a 10% additional capacity is factored in is ___ kVA.

_____ 8. If the oil heater (line 1) is added into the control circuit (option 4) and the heater is rated at 450 W, the new required kVA capacity of the transformer after a 10% additional capacity is factored in is ___ kVA.

Electric Motors

Review Questions

16

Name _____ Date _____

True-False

T   F   **1.** Work is any cause that changes the position, motion, direction, or shape of an object.

T   F   **2.** Electrical power is rated in horsepower or watts.

T   F   **3.** Full-load torque is the torque required to produce the rated power at full speed of the motor.

T   F   **4.** An induction motor is a motor that has no physical electrical connection to the rotor.

T   F   **5.** A DC series motor is a DC motor that has the armature and field circuits wired in parallel.

T   F   **6.** The stator is the rotating part of an AC motor.

T   F   **7.** An AC shunt motor is a motor that has the field connected in parallel with the armature.

T   F   **8.** Insulation breakdown is the main cause of motor failure.

Multiple Choice

_____ **1.** A DC ___ motor is a DC motor that has the series field connected in series with the armature and the shunt field connected in parallel with the series field and armature.
    A. permanent-magnet
    B. compound
    C. shunt
    D. series

_____ **2.** A DC ___ motor is a DC motor that has the series field coils connected in series with the armature.
    A. permanent
    B. compound
    C. shunt
    D. series

_____ **3.** A DC ___ motor is a DC motor that has only armature connections and no field connections.
    A. permanent magnet
    B. compound
    C. shunt
    D. series

_____ **4.** ___ torque is the torque required to start a motor.
    A. Locked rotor
    B. Pull-up
    C. Starting
    D. Full-load

_____ **5.** ___ torque is the force that produces or tends to produce rotation in a motor.
    A. Motor
    B. Shunt
    C. Series
    D. Pull-up

Completion

_____ **1.** A(n) ___ is a rotating output device that converts electrical power into a rotating, mechanical force.

_____ **2.** ___ is the force that causes an object to rotate.

_____ **3.** ___ torque is the torque a motor produces when the rotor is stationary and full power is applied to the motor.

_____ **4.** ___ torque is the torque required to bring a load up to its rated speed.

_____ **5.** ___ is the theoretical speed of a motor based on the motor's number of poles and the line frequency.

_____ **6.** A(n) ___ motor is an electric 1φ, AC motor that includes a running winding (main winding) and a starting winding (auxiliary winding).

_____ **7.** A(n) ___ switch is a switch that operates its contacts when a preset temperature is reached.

_____ **8.** ___ torque is the maximum torque a motor can produce without an abrupt reduction in motor speed.

_____ **9.** A(n) ___ motor is a 1φ, AC motor that includes a capacitor in addition to the running and starting windings.

_____ **10.** The ___ is the stationary part of an AC motor.

_____ **11.** A(n) ___ motor is a capacitor motor that has the starting winding and capacitor connected in series at all times.

_____ **12.** A(n) ___ motor is a motor that operates at more than one voltage level.

_____ **13.** ___ is the rate of doing work or using energy.

_____ **14.** ___ is a measure of the effectiveness with which a motor converts electrical energy to mechanical energy.

_____ **15.** ___ is the temperature of the air surrounding a piece of equipment.

Electric Motors

_____ **1.** Capacitor-start

_____ **2.** Split-phase

_____ **3.** DC shunt

_____ **4.** DC compound

_____ **5.** DC series

_____ **6.** Capacitor start-and-run

_____ **7.** 3φ

_____ **8.** Capacitor-run

Name _____ Date _____

Split-Phase Motors

1. Connect the motor to run in the clockwise direction on low voltage. Interconnect the motor terminals based on the motor nameplate. Connect the motor (motor terminals) to the starter (starter heater terminals) using only the required number of heaters and power contacts. Connect the starter (starter terminals) to the fused disconnect (fuse terminals) using only the required number of fuses. Connect the fused disconnect (disconnect terminals) to the best power supply (wye, delta, or DC) for the motor requirements. Run all wiring inside the conduit.

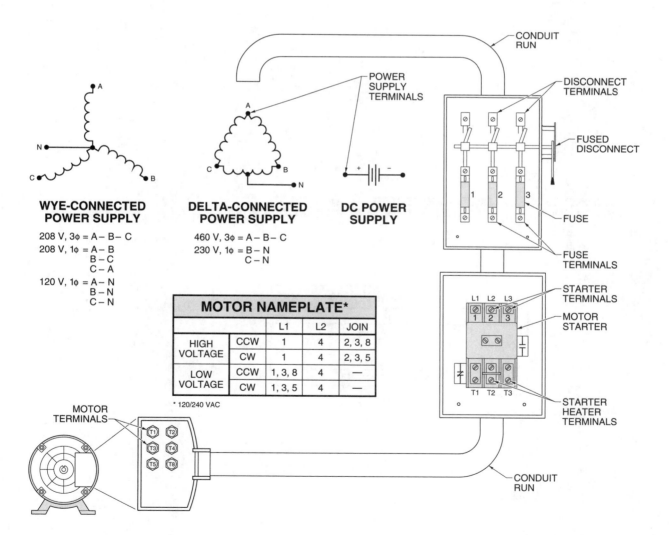

WYE-CONNECTED POWER SUPPLY

208 V, 3φ = A – B – C
208 V, 1φ = A – B
B – C
C – A
120 V, 1φ = A – N
B – N
C – N

DELTA-CONNECTED POWER SUPPLY

460 V, 3φ = A – B – C
230 V, 1φ = B – N
C – N

DC POWER SUPPLY

MOTOR NAMEPLATE*		L1	L2	JOIN
HIGH VOLTAGE	CCW	1	4	2, 3, 8
	CW	1	4	2, 3, 5
LOW VOLTAGE	CCW	1, 3, 8	4	—
	CW	1, 3, 5	4	—

* 120/240 VAC

Name _____ **Date** _____

Capacitor Motors

1. Connect the motor to run in the counterclockwise direction on high voltage. Interconnect the motor terminals based on the motor nameplate. Connect the motor (motor terminals) to the starter (starter heater terminals) using only the required number of heaters and power contacts. Connect the starter (starter terminals) to the fused disconnect (fuse terminals) using only the required number of fuses. Connect the fused disconnect (disconnect terminals) to the best power supply (wye, delta, or DC) for the motor requirements. Run all wiring inside the conduit.

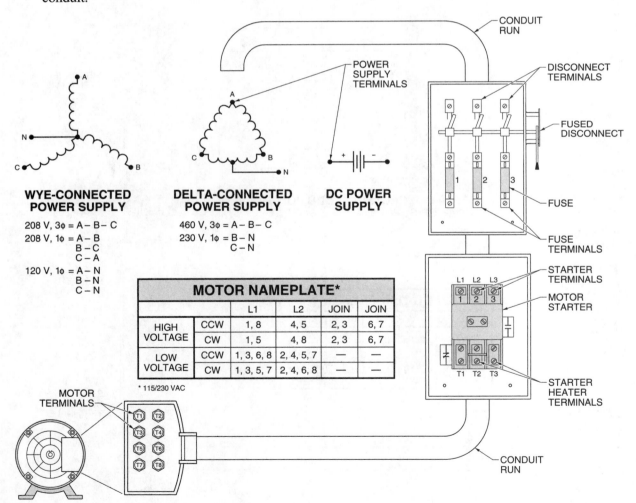

WYE-CONNECTED POWER SUPPLY

208 V, 3φ = A – B – C
208 V, 1φ = A – B
            B – C
            C – A
120 V, 1φ = A – N
            B – N
            C – N

DELTA-CONNECTED POWER SUPPLY

460 V, 3φ = A – B – C
230 V, 1φ = B – N
            C – N

DC POWER SUPPLY

MOTOR NAMEPLATE*		L1	L2	JOIN	JOIN
HIGH VOLTAGE	CCW	1, 8	4, 5	2, 3	6, 7
	CW	1, 5	4, 8	2, 3	6, 7
LOW VOLTAGE	CCW	1, 3, 6, 8	2, 4, 5, 7	—	—
	CW	1, 3, 5, 7	2, 4, 6, 8	—	—

* 115/230 VAC

Name _____ Date _____

Three-Phase Wye-Connected Motors

1. Connect the motor to run in the clockwise direction on low voltage. Interconnect the motor terminals based on the motor nameplate. Connect the motor (motor terminals) to the starter (starter heater terminals) using only the required number of heaters and power contacts. Connect the starter (starter terminals) to the fused disconnect (fuse terminals) using only the required number of fuses. Connect the fused disconnect (disconnect terminals) to the best power supply (wye, delta, or DC) for the motor requirements. Run all wiring inside the conduit.

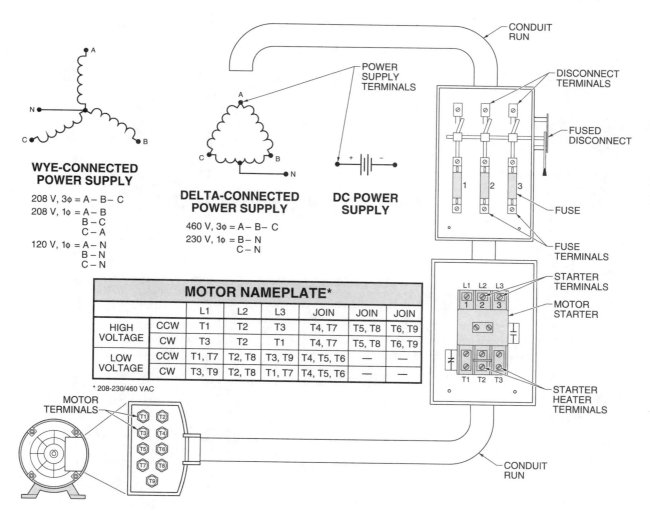

WYE-CONNECTED POWER SUPPLY

208 V, 3φ = A – B – C
208 V, 1φ = A – B
           B – C
           C – A
120 V, 1φ = A – N
           B – N
           C – N

DELTA-CONNECTED POWER SUPPLY

460 V, 3φ = A – B – C
230 V, 1φ = B – N
           C – N

DC POWER SUPPLY

MOTOR NAMEPLATE*		L1	L2	L3	JOIN	JOIN	JOIN
HIGH VOLTAGE	CCW	T1	T2	T3	T4, T7	T5, T8	T6, T9
	CW	T3	T2	T1	T4, T7	T5, T8	T6, T9
LOW VOLTAGE	CCW	T1, T7	T2, T8	T3, T9	T4, T5, T6	—	—
	CW	T3, T9	T2, T8	T1, T7	T4, T5, T6	—	—

* 208-230/460 VAC

Name _____ Date _____

Three-Phase High Voltage Delta-Connected Motors

1. Connect the motor to run in the clockwise direction on high voltage. Interconnect the motor terminals based on the motor nameplate. Connect the motor (motor terminals) to the starter (starter heater terminals) using only the required number of heaters and power contacts. Connect the starter (starter terminals) to the fused disconnect (fuse terminals) using only the required number of fuses. Connect the fused disconnect (disconnect terminals) to the best power supply (wye, delta, or DC) for the motor requirements. Run all wiring inside the conduit.

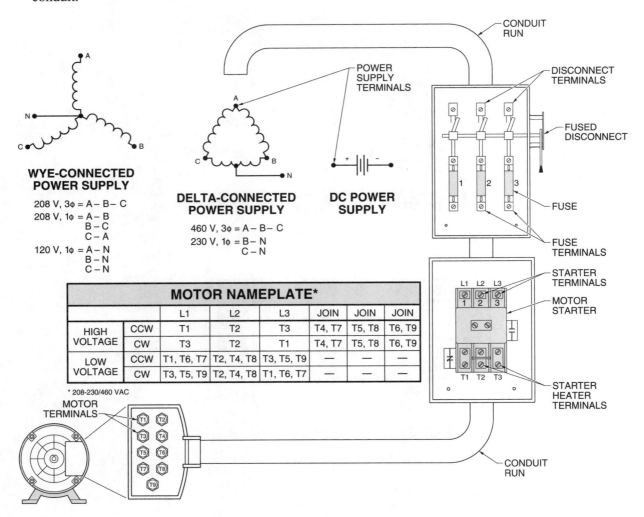

WYE-CONNECTED POWER SUPPLY

208 V, 3φ = A – B – C
208 V, 1φ = A – B
          B – C
          C – A
120 V, 1φ = A – N
          B – N
          C – N

DELTA-CONNECTED POWER SUPPLY

460 V, 3φ = A – B – C
230 V, 1φ = B – N
          C – N

DC POWER SUPPLY

MOTOR NAMEPLATE*		L1	L2	L3	JOIN	JOIN	JOIN
HIGH VOLTAGE	CCW	T1	T2	T3	T4, T7	T5, T8	T6, T9
	CW	T3	T2	T1	T4, T7	T5, T8	T6, T9
LOW VOLTAGE	CCW	T1, T6, T7	T2, T4, T8	T3, T5, T9	—	—	—
	CW	T3, T5, T9	T2, T4, T8	T1, T6, T7	—	—	—

* 208-230/460 VAC

Name _____ Date _____

Three-Phase Low Voltage Delta-Connected Motors

1. Connect the motor to run in the counterclockwise direction on high voltage. Interconnect the motor terminals based on the motor nameplate. Connect the motor (motor terminals) to the starter (starter heater terminals) using only the required number of heaters and power contacts. Connect the starter (starter terminals) to the fused disconnect (fuse terminals) using only the required number of fuses. Connect the fused disconnect (disconnect terminals) to the best power supply (wye, delta, or DC) for the motor requirements. Run all wiring inside the conduit.

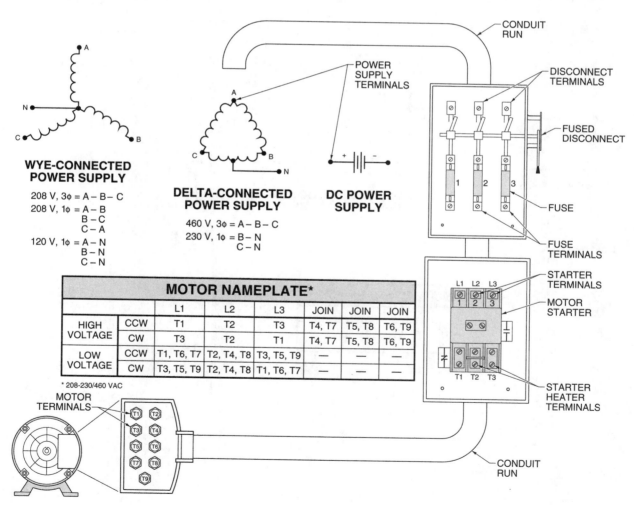

WYE-CONNECTED POWER SUPPLY

208 V, 3φ = A – B – C
208 V, 1φ = A – B
            B – C
            C – A
120 V, 1φ = A – N
            B – N
            C – N

DELTA-CONNECTED POWER SUPPLY

460 V, 3φ = A – B – C
230 V, 1φ = B – N
            C – N

DC POWER SUPPLY

MOTOR NAMEPLATE*							
		L1	L2	L3	JOIN	JOIN	JOIN
HIGH VOLTAGE	CCW	T1	T2	T3	T4, T7	T5, T8	T6, T9
	CW	T3	T2	T1	T4, T7	T5, T8	T6, T9
LOW VOLTAGE	CCW	T1, T6, T7	T2, T4, T8	T3, T5, T9	—	—	—
	CW	T3, T5, T9	T2, T4, T8	T1, T6, T7	—	—	—

* 208-230/460 VAC

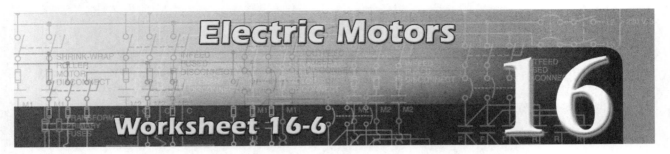

Name _____ **Date** _____

DC Motors

1. Connect the motor to run in the clockwise direction. Interconnect the motor terminals based on the motor nameplate. Connect the motor (motor terminals) to the starter (starter heater terminals) using only the required number of heaters and power contacts. Connect the starter (starter terminals) to the fused disconnect (fuse terminals) using only the required number of fuses. Connect the fused disconnect (disconnect terminals) to the best power supply (wye, delta, or DC) for the motor requirements. Run all wiring inside the conduit.

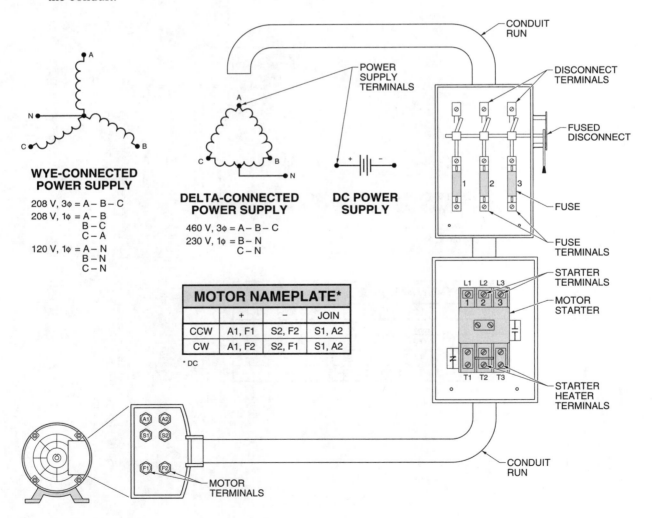

WYE-CONNECTED POWER SUPPLY

208 V, 3φ = A – B – C
208 V, 1φ = A – B
             B – C
             C – A
120 V, 1φ = A – N
             B – N
             C – N

DELTA-CONNECTED POWER SUPPLY

460 V, 3φ = A – B – C
230 V, 1φ = B – N
             C – N

DC POWER SUPPLY

MOTOR NAMEPLATE*			
	+	–	JOIN
CCW	A1, F1	S2, F2	S1, A2
CW	A1, F2	S2, F1	S1, A2

* DC

Name _____ Date _____

Measurement Activity Objectives

- Measure the locked rotor current of a motor.
- Measure the full-load current of a motor.

Procedure 16-1

Motor torque is the force that produces or tends to produce rotation in a motor. Motor specifications list the different torque characteristics of different motor types. Motor torque includes locked rotor torque, pull-up torque, full-load torque, and breakdown torque. Set a DMM to measure in-line mA DC current and connect the motor (fan) to the battery with the DMM connected in-line. Using the eraser of a pencil, stop the motor shaft from rotating.

_____ **1.** When the motor is connected to the battery and the motor shaft is not rotating, the measured current is ___ mA.

_____ **2.** With the motor shaft stopped, the torque that the motor is producing to attempt to rotate the shaft is ___ torque.

Remove the eraser from the motor shaft and allow the motor to reach full speed.

_____ **3.** When the motor shaft is rotating, the measured current is ___ mA.

_____ **4.** With the motor shaft rotating at full speed, the torque that the motor is producing to keep the shaft rotating is ___ torque.

Once again, stop the motor shaft from rotating using the eraser of a pencil. Watching the current reading on the DMM, slowly allow the motor to start rotating and reach full speed.

_____ **5.** As the motor shaft approaches full speed, does the measured current increase or decrease?

_____ **6.** As the motor shaft approaches full speed, the torque that the motor is producing during the acceleration of the shaft is ___ torque.

With the motor rotating at full speed, reduce the speed of the motor shaft by applying pressure with the eraser of a pencil and observe the current reading on the DMM.

_____ **7.** As the motor shaft speed decreases, does the measured current increase or decrease?

_____ **8.** As the motor shaft speed decreases, the torque that the motor is producing during the deceleration of the shaft is ___ torque.

Resistance, Inductance, and Capacitance

Review Questions

17

Name _____ Date _____

True-False

T   F   **1.** Conductor resistance is kept to a minimum by limiting the temperature in the circuit and by using the correct size, length, and material of the conductor.

T   F   **2.** Capacitance is the property of a circuit that causes it to oppose a change in current due to energy stored in a magnetic field.

T   F   **3.** The total inductance in a circuit containing parallel-connected coils is greater than the smallest coil value.

T   F   **4.** As the area of the conductor increases, its resistance decreases.

T   F   **5.** As the length of the conductor (current path) increases, its resistance decreases.

T   F   **6.** The amount of inductance produced by a coil depends on the strength of the magnetic field produced by the coil.

Multiple Choice

_____ **1.** Inductors are generally added into a circuit to control the amount of ___ flowing in a circuit or a branch of the circuit.
     A. voltage
     B. resistance
     C. pressure
     D. current

_____ **2.** Fixed capacitors include ___.
     A. mica
     B. paper
     C. ceramic
     D. all of the above

_____ **3.** ___-core inductors consist of a coil (copper wire) wrapped around a form (plastic, ceramic, or Bakelite) with no material in the middle.
     A. Air
     B. Iron
     C. Ferrite
     D. Copper

171

_____ **4.** ___ resistance is the actual (true) resistance of a component when operating current is passing through the device.
    A. Cold
    B. Warm
    C. Hot
    D. Inductive

_____ **5.** A flashlight is an example of a simple electric circuit and includes a ___.
    A. power supply (batteries)
    B. switch
    C. load (lamp)
    D. all of the above

Completion

_____ **1.** A(n) ___ is any device that converts electrical energy to mechanical energy, heat, light, or sound.

_____ **2.** A(n) ___ circuit is a circuit that contains only resistance.

_____ **3.** A(n) ___ is a material that has very little resistance and permits electrons to move through it easily.

_____ **4.** ___ is the resistance of a conductor having a specific length and cross-sectional area.

_____ **5.** ___ is an inductor's opposition to alternating current.

_____ **6.** ___ is the ability of a component or circuit to store energy in the form of an electrical charge.

_____ **7.** ___ resistance is the resistance of a component when operating current is not passing through the device.

_____ **8.** A(n) ___ capacitor is a capacitor that has one value of capacitance.

_____ **9.** ___ is the opposition to current flow by a capacitor.

_____ **10.** ___ is the total opposition of any combination of resistance, inductive reactance, and capacitive reactance offered to the flow of alternating current.

Capacitors

_____ **1.** Mica

_____ **2.** Electrolytic

_____ **3.** Ceramic

_____ **4.** Variable

_____ **5.** Paper (plastic)

A   **B**

POLARITY
MARK

C   **D**   **E**

Name _____ **Date** _____

Inductive Reactance

_____ **1.** $X_L = $ ___ $k\Omega$

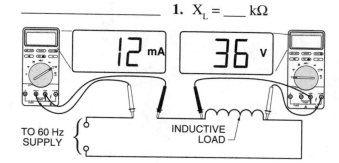

TO 60 Hz SUPPLY

12 mA   36 V

INDUCTIVE LOAD

_____ **2.** $X_L = $ ___ $k\Omega$

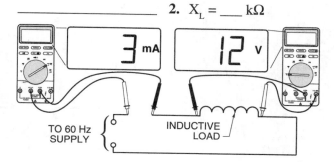

TO 60 Hz SUPPLY

3 mA   12 V

INDUCTIVE LOAD

_____ **3.** $X_L = $ ___ Ω

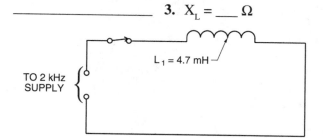

TO 2 kHz SUPPLY

$L_1 = 4.7$ mH

_____ **4.** $X_L = $ ___ $k\Omega$

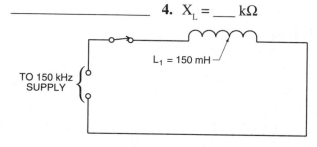

TO 150 kHz SUPPLY

$L_1 = 150$ mH

_____ **5.** $L_T = $ ___ mH

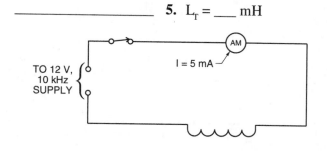

TO 12 V, 10 kHz SUPPLY

AM

$I = 5$ mA

_____ **6.** $L_T = $ ___ μH

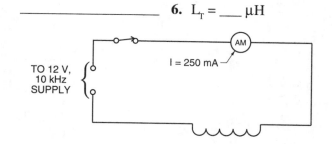

TO 12 V, 10 kHz SUPPLY

AM

$I = 250$ mA

_____ **7.** $f = $ ___ Hz

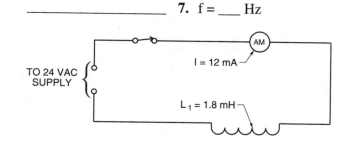

TO 24 VAC SUPPLY

AM

$I = 12$ mA

$L_1 = 1.8$ mH

_____ **8.** $f = $ ___ Hz

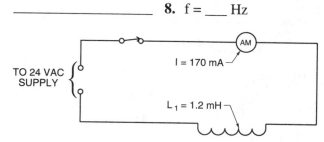

TO 24 VAC SUPPLY

AM

$I = 170$ mA

$L_1 = 1.2$ mH

Name _____ Date _____

Series Inductive Reactance

_____ **1.** L_T = ___ mH

_____ **2.** X_L = ___ Ω

_____ **3.** Meter 3 reads ___ mA

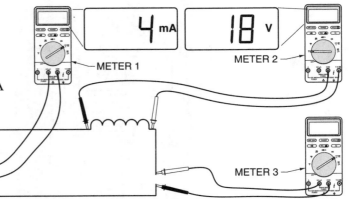

METER 1 · METER 2 · METER 3 · TO 1000 Hz SUPPLY

_____ **4.** L_T = ___ μH

TO 1000 Hz SUPPLY
$L_1 = 22$ μH
$L_3 = 50$ mH $L_2 = 100$ μH

_____ **5.** L_T = ___ μH

TO 20 kHz SUPPLY
$L_1 = 68$ μH
$L_2 = 20$ mH

_____ **6.** X_L = ___ kΩ

TO 100 kHz SUPPLY
$L_1 = 12$ μH
$L_3 = 1000$ μH $L_2 = 100$ μH

_____ **7.** X_L = ___ kΩ

TO 180 kHz SUPPLY
$L_1 = 8$ mH
$L_2 = 8$ μH

_____ **8.** f = ___ Hz

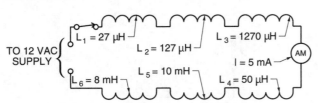

TO 12 VAC SUPPLY
$L_1 = 27$ μH $L_3 = 1270$ μH
$L_2 = 127$ μH
I = 5 mA
$L_6 = 8$ mH $L_5 = 10$ mH $L_4 = 50$ μH

_____ **9.** f = ___ Hz

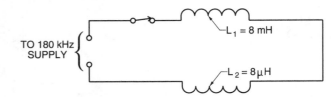

TO 12 VAC SUPPLY
$L_1 = 80$ μH $L_2 = 150$ μH $L_3 = 72$ μH
I = 5 mA
$L_6 = 12$ μH $L_5 = 12,000$ μH $L_4 = 50$ μH

Name _____ Date _____

Parallel Inductive Reactance

_____ **1.** $L_T =$ ___ mH

_____ **2.** $X_L =$ ___ Ω

_____ **3.** Meter 3 reads ___ mA

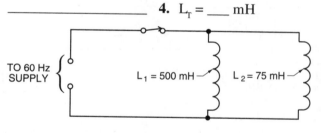

TO 60 Hz SUPPLY — $L_1 = 6$ mH, $L_2 = 3$ mH, $L_3 = 2$ mH, METER 1 **1.59** A, METER 2 **10** V, METER 3

_____ **4.** $L_T =$ ___ mH

TO 60 Hz SUPPLY — $L_1 = 500$ mH, $L_2 = 75$ mH

_____ **5.** $L_T =$ ___ mH

TO 60 Hz SUPPLY — $L_1 = 5$ mH, $L_2 = 20$ mH, $L_3 = 10$ mH

_____ **6.** $X_L =$ ___ Ω

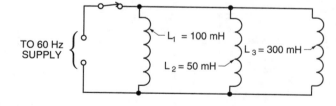

TO 60 Hz SUPPLY — $L_1 = 100$ mH, $L_2 = 50$ mH, $L_3 = 300$ mH

_____ **7.** $X_L =$ ___ Ω

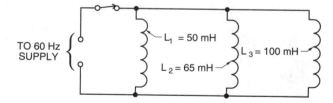

TO 60 Hz SUPPLY — $L_1 = 50$ mH, $L_2 = 65$ mH, $L_3 = 100$ mH

_____ **8.** f = ___ Hz

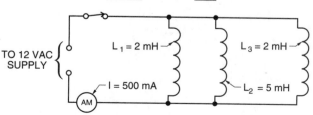

TO 12 VAC SUPPLY — $L_1 = 2$ mH, $L_3 = 2$ mH, $L_2 = 5$ mH, I = 500 mA, AM

_____ **9.** f = ___ Hz

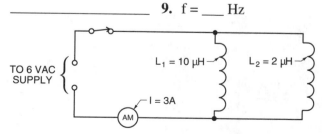

TO 6 VAC SUPPLY — $L_1 = 10$ μH, $L_2 = 2$ μH, I = 3A, AM

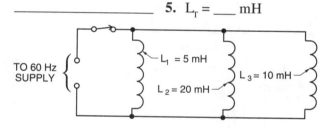

Name _____ **Date** _____

Capacitive Reactance

_____ **1.** $X_C =$ ___ Ω

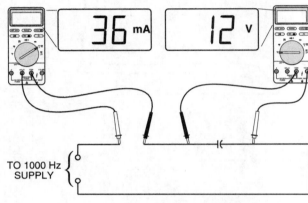

_____ **2.** $X_C =$ ___ Ω

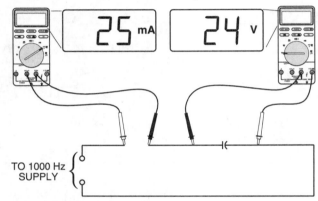

_____ **3.** $X_C =$ ___ Ω

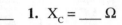

TO 60 Hz SUPPLY

$C_1 = 90\ \mu F$

_____ **4.** $X_C =$ ___ Ω

TO 1000 Hz SUPPLY

$C_1 = 90\ \mu F$

_____ **5.** $X_C =$ ___ Ω

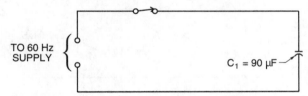

TO 5 kHz SUPPLY

$C_1 = 12\ \mu F$

_____ **6.** $X_C =$ ___ Ω

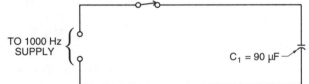

TO 5 kHz SUPPLY

$C_1 = 24\ \mu F$

_____ **7.** $X_C =$ ___ Ω

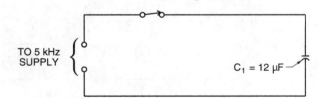

TO 10 kHz SUPPLY

$C_1 = 50\ \mu F$

_____ **8.** $X_C =$ ___ Ω

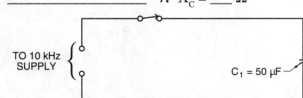

TO 10 kHz SUPPLY

$C_1 = 5\ \mu F$

Name _____ **Date** _____

Series Capacitive Reactance

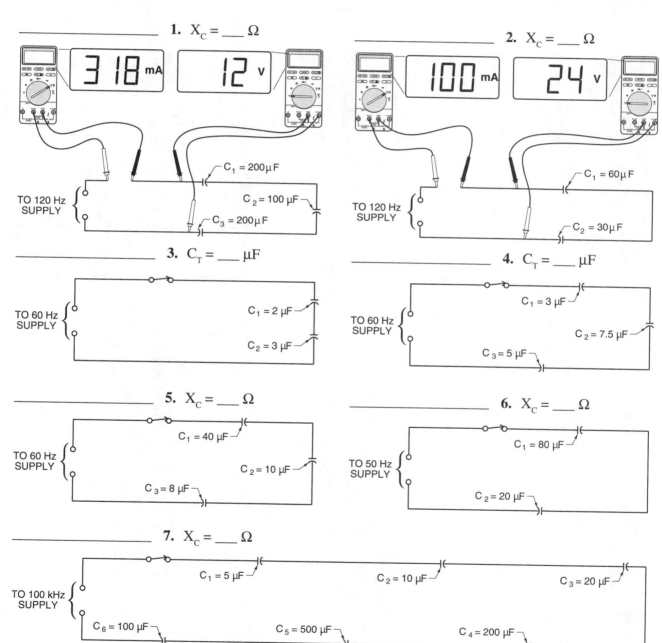

1. $X_C =$ ___ Ω

318 mA   12 V

TO 120 Hz SUPPLY
$C_1 = 200\,\mu F$
$C_2 = 100\,\mu F$
$C_3 = 200\,\mu F$

2. $X_C =$ ___ Ω

100 mA   24 V

TO 120 Hz SUPPLY
$C_1 = 60\,\mu F$
$C_2 = 30\,\mu F$

3. $C_T =$ ___ μF

TO 60 Hz SUPPLY
$C_1 = 2\,\mu F$
$C_2 = 3\,\mu F$

4. $C_T =$ ___ μF

TO 60 Hz SUPPLY
$C_1 = 3\,\mu F$
$C_2 = 7.5\,\mu F$
$C_3 = 5\,\mu F$

5. $X_C =$ ___ Ω

TO 60 Hz SUPPLY
$C_1 = 40\,\mu F$
$C_2 = 10\,\mu F$
$C_3 = 8\,\mu F$

6. $X_C =$ ___ Ω

TO 50 Hz SUPPLY
$C_1 = 80\,\mu F$
$C_2 = 20\,\mu F$

7. $X_C =$ ___ Ω

TO 100 kHz SUPPLY
$C_1 = 5\,\mu F$
$C_2 = 10\,\mu F$
$C_3 = 20\,\mu F$
$C_6 = 100\,\mu F$
$C_5 = 500\,\mu F$
$C_4 = 200\,\mu F$

177

Name _____ **Date** _____

Parallel Capacitive Reactance

_____ **1.** $C_T =$ ___ μF   _____ **3.** $C_T =$ ___ μF

_____ **2.** $X_C =$ ___ Ω   _____ **4.** $X_C =$ ___ Ω

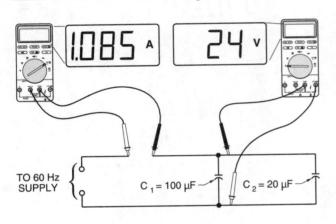

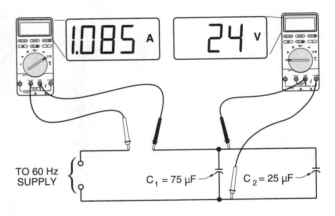

_____ **5.** $C_T =$ ___ μF   _____ **6.** $C_T =$ ___ μF

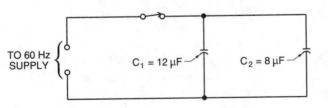

_____ **7.** $X_C =$ ___ Ω   _____ **8.** $X_C =$ ___ Ω

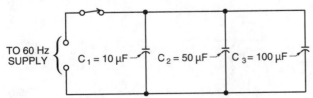

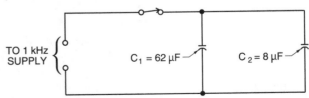

Name _____ **Date** _____

Measurement Activity Objectives
- Build a circuit to verify that there is no inductive reactance in a DC circuit.
- Measure resistance, voltage, and current in a circuit that includes a coil and is DC powered.

Procedure 17-1

In DC circuits (circuits that have zero frequency), there is no opposition due to inductance. In DC circuits, a coil has low resistance and no inductive reactance. Thus, only the resistance of the wire limits current flow through a coil connected to DC.

_____ **1.** The measured resistance of the coil on the relay included in the component kit is ___ Ω.

_____ **2.** The measured voltage of the battery is ___ VDC.

Using the measured resistance and voltage, apply Ohm's law to calculate the amount of current based on only the resistance of the coil wire.

_____ **3.** The calculated current is ___ mA.

Build the following circuit using the components provided.

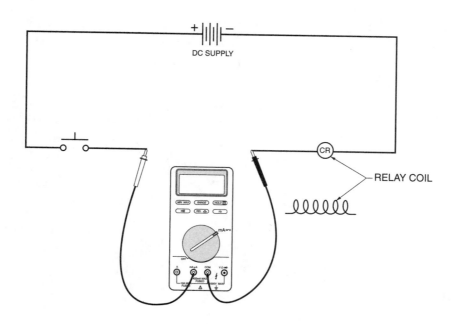

Test the circuit by closing the switch. The relay coil should energize. Set a DMM to take an in-line DC mA current measurement. Open the circuit and measure the current of the coil.

_____ **4.** The current of the coil is ___ mA DC.

5. Is the measured current equal to the calculated current? Why or why not?

Name _____ Date _____

True-False

T   F   **1.** A network is a system of computers, terminals, and databases connected by communication lines.

T   F   **2.** Any change in the power source voltage affects the load's operation.

T   F   **3.** An AC motor should generally operate within ±20% of its voltage rating.

T   F   **4.** Transverse noise is noise produced between the ground and hot or the ground and neutral lines.

T   F   **5.** A load is any device that converts electrical energy into light, heat, motion, or sound.

Multiple Choice

_____ **1.** A ___ is temporary, unwanted voltage in an electrical circuit.
A. voltage stabilizer
B. transient voltage
C. voltage meter
D. transverse voltage

_____ **2.** A ___ is the number of closed contact positions per pole.
A. break
B. throw
C. pole
D. solid-state relay

_____ **3.** A(n) ___ is an input that sends a continuously changing variable into the system.
A. network
B. normally closed input
C. analog input
D. input/output network

_____ 4. ___ lumen is the amount of light produced when a lamp is new.
   A. Mean
   B. Load
   C. Initial
   D. Manufactured

_____ 5. ___ noise is noise produced between the ground and hot or the ground and neutral lines.
   A. Electrical
   B. Common
   C. Transverse
   D. Ground

Completion

_____ 1. To produce work, electrical circuits must include a(n) ___, a source of electricity, and a method of controlling the flow of electricity.

_____ 2. A(n) ___ is a combination of components interconnected to perform work or meet a specific need.

_____ 3. A(n) ___ is a device that is used to produce work, light, heat, sound, or display information.

_____ 4. A voltage ___ is a device that provides precise voltage regulation to protect equipment from voltage dips and voltage surges.

_____ 5. A(n) ___ is the deliberate reduction of the voltage level by a utility company to conserve energy during peak usage times.

_____ 6. A voltage ___ is a momentary low voltage.

_____ 7. ___ is any unwanted disturbance that interferes with a signal.

_____ 8. A voltage ___ is a higher-than-normal voltage that temporarily exists on one or more power lines.

_____ 9. A(n) ___ is a device that produces electricity when two different metals (such as iron and constantan) that are joined together are heated.

_____ 10. ___ is the difference in voltage between a voltage surge and a voltage dip.

_____ 11. A(n) ___ is a device that converts solar energy to electrical energy.

_____ 12. A(n) ___ relay is a switching device that has sets of contacts which are closed by a magnetic force.

_____ 13. A(n) ___ is a control device that uses a small control current to energize or de-energize the load connected to it.

_____ 14. ___ noise is noise produced between the hot and neutral lines.

_____ 15. A(n) ___ relay is a relay that uses electronic switching devices in place of mechanical contacts.

Circuit Requirements

Worksheet 18-1

18

Name _____ **Date** _____

Heating Element Voltage Variations

Applied Voltage**	PERCENT RATED POWER*										
	Rated Voltage***										
	110	115	120	208	220	230	240	277	440	460	480
110	100	91	84	28	25	23	21	16	6.2	5.7	5.2
115	109	100	92	31	27	25	23	17	6.7	6.2	5.7
120	119	109	100	33	30	27	25	19	7.4	6.8	6.3
208			300	100	89	82	75	56	22	20	19
220				112	100	91	84	63	25	23	21
230				122	109	109	92	69	27	25	23
240				133	119		100	75	30	27	25
277							133	100	40	36	33
380								188	74	68	63
415									89	81	75
440									100	91	84
460									109	100	92
480									119	109	100

* in W
** in V
*** in %

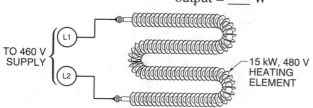

_____ **1.** Heating element true output = ___ W

TO 460 V SUPPLY L1 L2

15 kW, 480 V HEATING ELEMENT

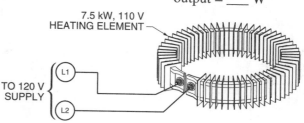

_____ **2.** Heating element true output = ___ W

7.5 kW, 110 V HEATING ELEMENT

TO 120 V SUPPLY L1 L2

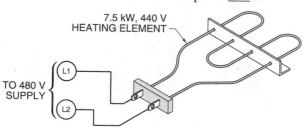

_____ **3.** Heating element true output = ___ W

7.5 kW, 440 V HEATING ELEMENT

TO 480 V SUPPLY L1 L2

_____ **4.** Heating element true output = ___ W

TO 208 V SUPPLY L1 L2

300 W, 220 V HEATING ELEMENT

183

Name _____ Date _____

Manufacturer's Lamp Characteristics Chart Use

METAL-HALIDE LAMP CHARACTERISTICS							
Lamp Wattage	Voltage Rating	Fuse*	Starting Current*	Operational Current*	Line Input Watts	% Allowable Variation	
						Operating Voltage Range	Lamp Wattage
50	120	3	.60	.65	72	A	±8
	277	3	.25	.30			
100	120	8	1.15	1.15	129	A	±12
	208	5	.66	.66			
	240	3	.58	.58			
	277	3	.50	.50			
175	120	5	1.30	1.80	210	B	±10
	208	3	.75	1.05			
	240	3	.65	.90			
	277	3	.55	.80			
	480	3	.35	.45			

* in A

_____ **1.** Operating current at A = ___ A

_____ **2.** Operating current at B = ___ A

_____ **3.** Operating current at C = ___ A

_____ **4.** Operating current at D = ___ A

VOLTAGE RANGES*	
A (5%)	B (10%)
126 – 120 – 114	132 – 120 – 108
218 – 208 – 198	229 – 208 – 187
252 – 240 – 228	264 – 240 – 216
291 – 277 – 263	305 – 277 – 249

* in V

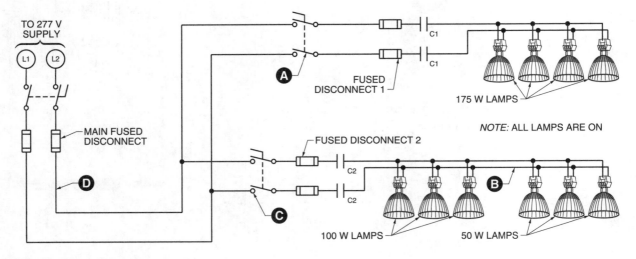

TO 277 V SUPPLY

L1 L2

MAIN FUSED DISCONNECT

FUSED DISCONNECT 1

175 W LAMPS

NOTE: ALL LAMPS ARE ON

FUSED DISCONNECT 2

100 W LAMPS   50 W LAMPS

Name _____ Date _____

Mechanical Relay Interface Application

1. Complete the wiring diagram from inside the relay control panel to the control switches using the line diagram.

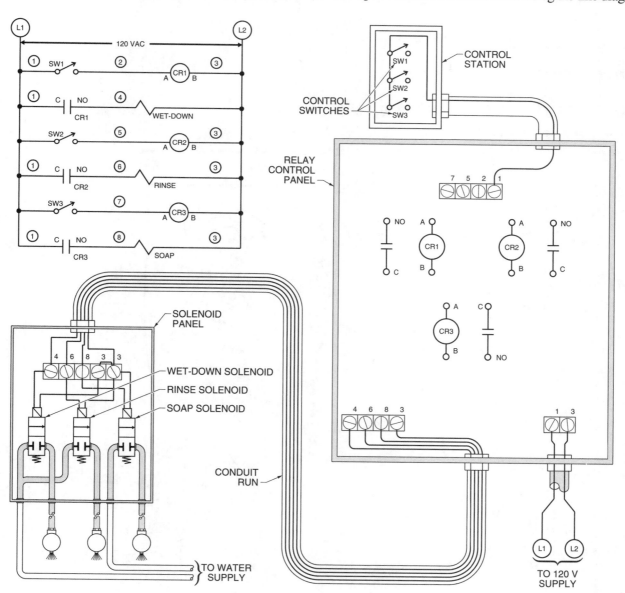

Name _____ Date _____

Solid-State Relay Interface Application

1. Complete the control circuit wiring diagram using the line diagram.

CONTROL CIRCUIT

12 VDC SUPPLY

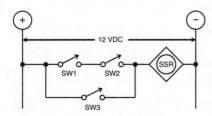

SW1

SW2

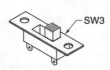

SW3

POWER CIRCUIT

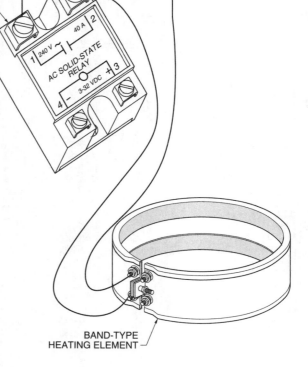

AC OUTPUT
SOLID-STATE RELAY

BAND-TYPE
HEATING ELEMENT

186

Name _____ **Date** _____

Contactor Interface Application

1. Complete the wiring diagram for the lighting contactor using the line diagram of the power circuit. *Note:* Not all contacts on the lighting contactor are used.

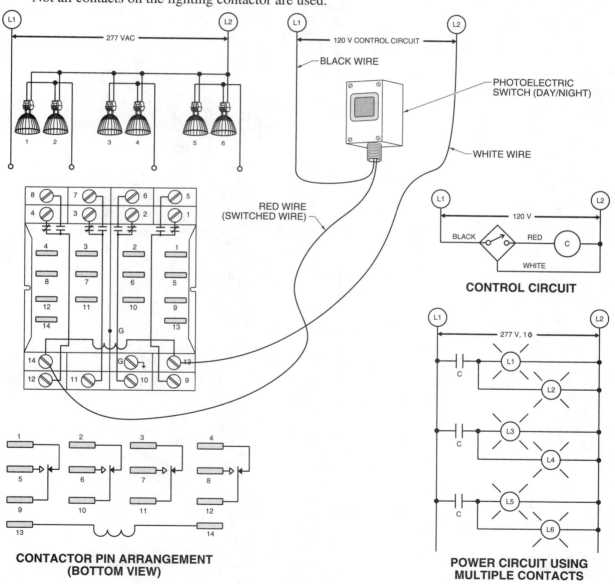

CONTACTOR PIN ARRANGEMENT (BOTTOM VIEW)

CONTROL CIRCUIT

POWER CIRCUIT USING MULTIPLE CONTACTS

Name _____ Date _____

Motor Starter Interface Application

1. Complete the wiring diagram using the line diagram.

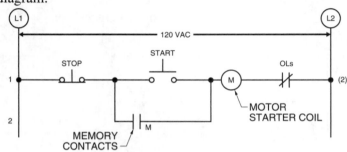

CONTROL CIRCUIT

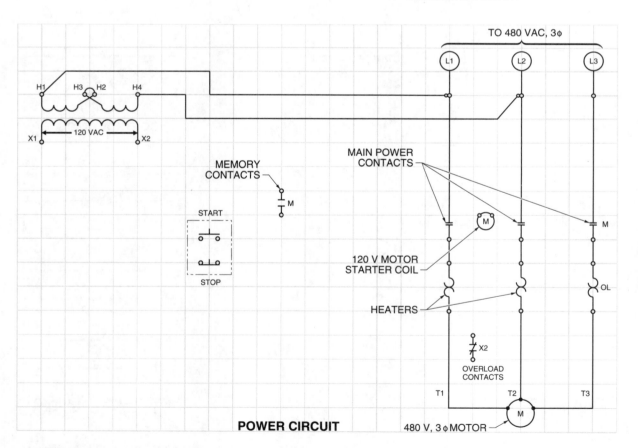

POWER CIRCUIT

Circuit Requirements

Activities

18

Name _____ **Date** _____

Drawing Activity Objectives

- Draw a circuit showing how a control relay (CR) is used as a circuit interface.
- Draw a circuit showing how a control relay is used to meet a circuit requirement for adding memory into a control circuit.

Procedure 18-1

Control relays are used as interfaces and for adding circuit memory as part of circuit requirements in many types of circuits.

1. Draw a circuit (Circuit 1) using standard electrical symbols that includes a 9 VDC power supply, a normally open pushbutton connected to control a control relay (CR) coil, a normally open relay contact that controls a red lamp, and a normally closed relay contact that controls a green lamp.

_____ 2. Is the red lamp or the green lamp ON before the pushbutton is pressed?

_____ 3. Is the red lamp or the green lamp ON when the pushbutton is pressed and held down?

4. Draw another circuit (Circuit 2) by adding a normally open relay contact and a normally closed limit switch in parallel with the pushbutton from Circuit 1. The normally open relay contact is in series with the input of the normally closed limit switch.

_____ **5.** Is the red lamp or the green lamp ON before the pushbutton is pressed?

_____ **6.** Is the red lamp or the green lamp ON when the pushbutton is pressed and released?

_____ **7.** What does the limit switch do in this circuit?

Hands-On Activity Objectives

• Build and test a circuit that uses a relay interface to multiply the number of output contacts of a switch.
• Build and test a circuit that uses a relay interface to develop circuit memory.

Procedure 18-2

Build Circuit 1 using the drawing in Procedure 18-1 and test it by pressing and releasing the pushbutton.

_____ **1.** Does the circuit operate as designed?

Build Circuit 2 using the drawing in Procedure 18-1 and test it by pressing and releasing the pushbutton.

_____ **2.** Does the red lamp turn ON and the green lamp turn OFF?

Press the limit switch.

_____ **3.** Do the lamps change back?

Add the motor in parallel with the red lamp (polarity is important) and test the circuit again.

_____ **4.** Does the motor stay ON when the pushbutton is pressed and released?

_____ **5.** Does the motor turn OFF when the limit switch is pressed and released?

Measurement Activity Objectives

• Build a test circuit on which to take electrical measurements.
• Measure the voltage of the power supply before and after it is loaded (loads ON).
• Measure the current of the lamp, relay coil, and motor.

Procedure 18-3

Build Circuit 2 using the drawing in Procedure 18-1 and answer the following:

_____ **1.** The measured voltage at the power supply (battery) before the pushbutton is pressed is ___ V.

_____ **2.** The measured voltage at the power supply after the pushbutton is pressed is ___ V.

  3. Was there a voltage difference? Why or why not?

_____ **4.** The measured in-line current of the green lamp when energized is ___ mA.

_____ **5.** The measured in-line current of the CR coil when energized is ___ mA.

_____ **6.** The measured in-line current of the motor when energized is ___ mA.

Residential Circuits

Review Questions

19

Name _____ **Date** _____

True-False

T　　F　　**1.** The three standard manually-operated switches used to control most ON/OFF lighting circuits are two-way, three-way, and four-way switches.

T　　F　　**2.** A split-wired receptacle is a standard receptacle that has had the tab between the two brass-colored (hot) terminal screws removed.

T　　F　　**3.** Electrical systems primarily used to convert raw materials and manufacture products are generally referred to as commercial systems.

T　　F　　**4.** Two-way switches have ON and OFF position markings on them because they have distinct ON and OFF positions.

T　　F　　**5.** Motion sensors are connected in the same manner as a two-way switch.

T　　F　　**6.** A four-way switch has one common terminal and two traveler terminals.

Multiple Choice

_____ **1.** A ___ is a mounting device that screws onto the threaded fixture stud and takes a threaded nipple.
- A. tap
- B. splice
- C. hanger
- D. hickey

_____ **2.** A ___ switch is a single-pole, single-throw (SPST) switch.
- A. two-way
- B. three-way
- C. four-way
- D. dimmer

_____ **3.** A ___ switch is a switch that changes the lamp brightness by changing the voltage applied to the lamp.
- A. toggle
- B. dimmer
- C. three-way
- D. four-way

191

_____ **4.** A ___ is a device used to connect equipment with a cord and plug to an electrical system.
- A. switch
- B. disconnect
- C. receptacle
- D. splice

_____ **5.** A ___ is a low-resistance conducting connection between electrical circuits, equipment, and the earth.
- A. disconnect
- B. ground
- C. fault
- D. fuse

_____ **6.** A ___ switch is a double-pole, double-throw switch that changes the electrical connections inside the switch from straight to diagonal.
- A. two-way
- B. three-way
- C. four-way
- D. dimmer

Completion

_____ **1.** A(n) ___ receptacle is a receptacle that has two spaces for connecting two different plugs.

_____ **2.** A(n) ___ receptacle is an electrical receptacle that minimizes electrical noise by providing a separate grounding path for each receptacle.

_____ **3.** A(n) ___ is any amount of current above the level that may deliver a dangerous shock.

_____ **4.** A(n) ___ is an electric device which protects personnel by detecting ground faults and quickly disconnecting power from the circuit.

_____ **5.** ___ is heat in the form of radiant energy.

_____ **6.** A(n) ___ is an infrared sensing device that detects the movement of a temperature variance and automatically switches when the movement is detected.

Receptacles

_____ **1.** GFCI

_____ **2.** Isolated ground

_____ **3.** Standard

ORANGE TRIANGLE AND/OR ORANGE FACE

Ⓐ   Ⓑ   Ⓒ

Name _____ Date _____

Two-Way and Three-Way Switch Lamp Control

1. Connect the devices based on the architectural plan.

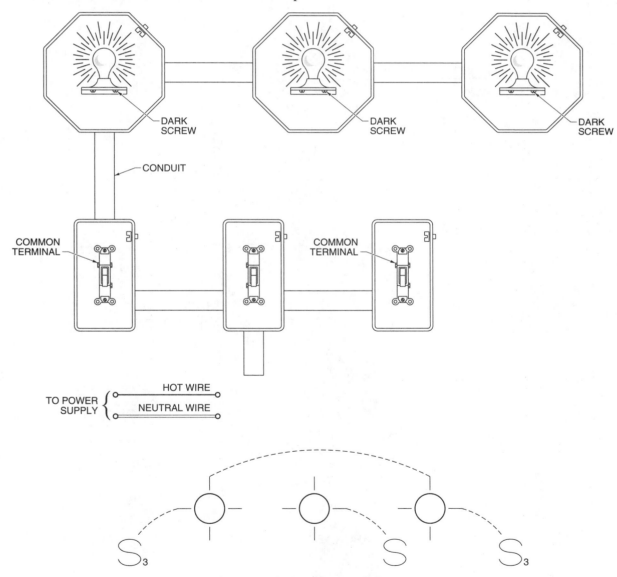

ARCHITECTURAL PLAN

Name _____ Date _____

Three Location Lamp Control

1. Connect the devices based on the architectural plan.

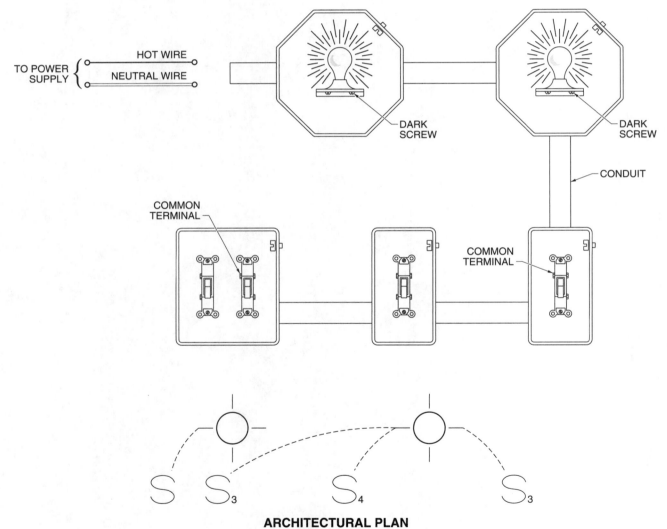

ARCHITECTURAL PLAN

Name _____ Date _____

Combination Lamp/Receptacle Circuit Wiring

1. Connect the devices based on the architectural plan.

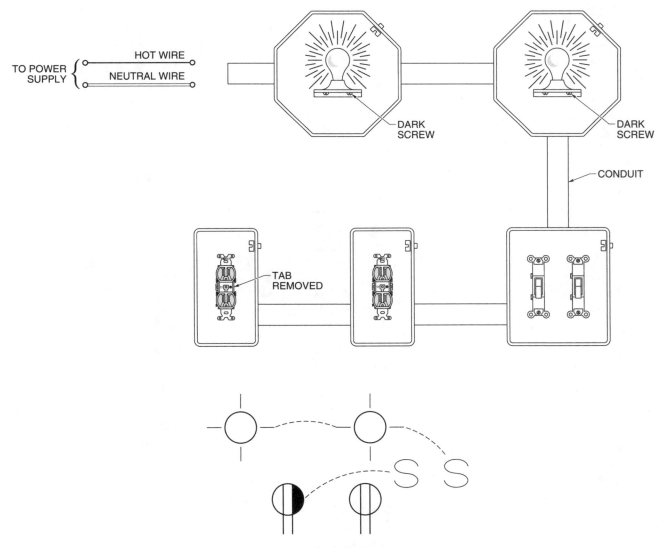

ARCHITECTURAL PLAN

Name _____ Date _____

Two Location Receptacle Control

1. Connect the devices based on the architectural plan.

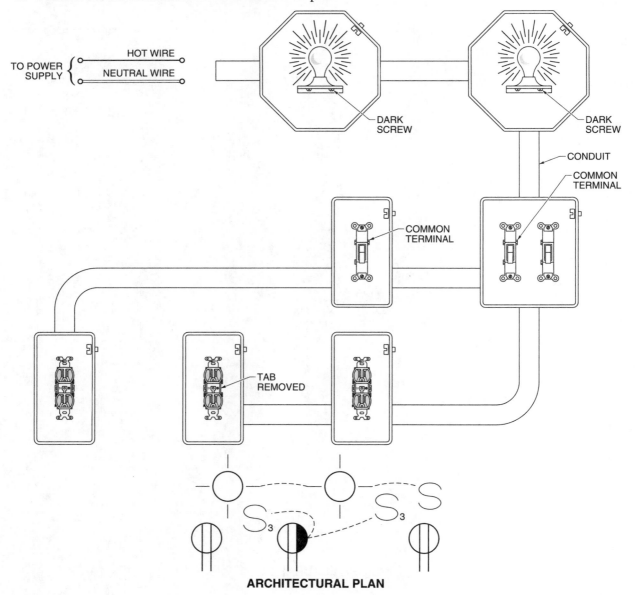

ARCHITECTURAL PLAN

196

Name _____ Date _____

Lamp/Receptacle Switch Control

1. Connect the devices based on the architectural plan.

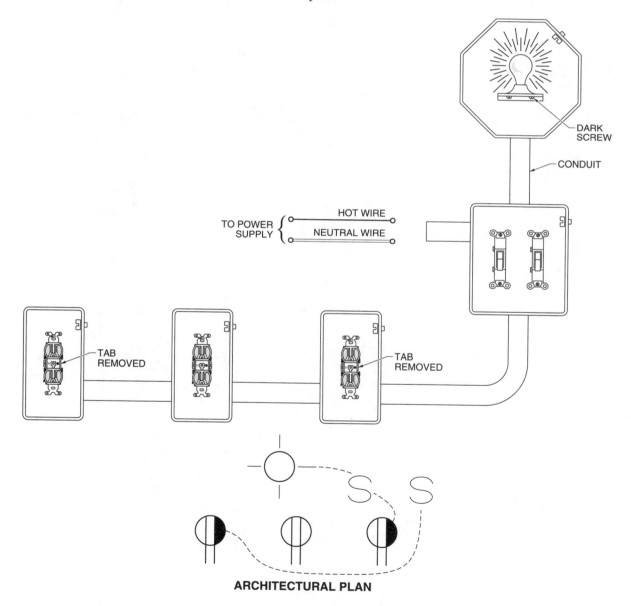

Name _____ Date _____

Combination Circuit Wiring

1. Connect the devices based on the architectural plan.

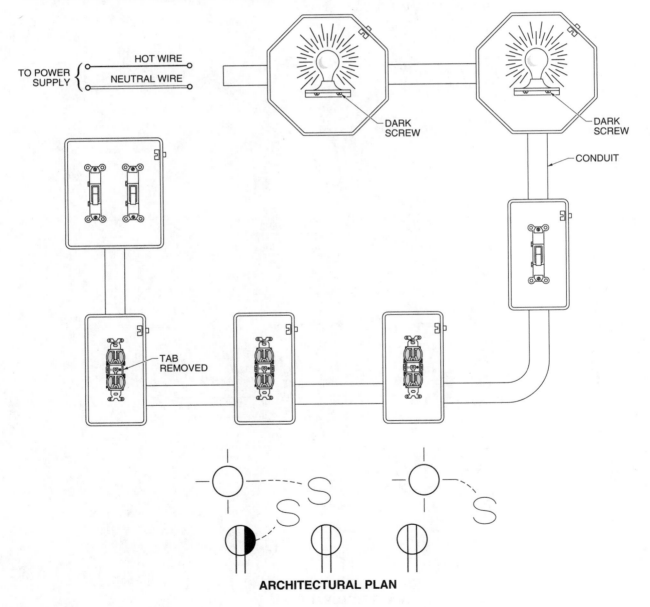

ARCHITECTURAL PLAN

Name _____ Date _____

Drawing Activity Objectives

- Complete a truth table for a circuit that uses two three-way switches to control a lamp (ON/OFF) from either of two switch locations.

- Draw an equivalent circuit, using an exclusive OR (XOR) logic block, that uses two two-way switches to control a lamp (ON/OFF) from either of two switch locations.

- Complete a truth table for a circuit that uses two three-way switches and one four-way switch to control a lamp (ON/OFF) from any of three switch locations.

- Draw an equivalent circuit, using two exclusive OR (XOR) logic blocks, that uses three two-way switches to control a lamp (ON/OFF) from any of the three switch locations.

Procedure 19-1

1. Using Circuit 1, complete the truth table for each switch position.

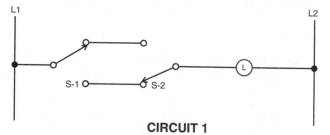

CIRCUIT 1

S-1	S-2	LAMP (ON or OFF)
UP	DOWN	
DOWN	DOWN	
DOWN	UP	
UP	UP	

2. Connect the two input switches (I-01 and I-02) to the input of the XOR logic block (B-01) and the output of the XOR logic block to the output (Q-01).

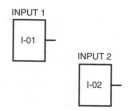

CIRCUIT 2

3. Using Circuit 2, complete the truth table for each switch position.

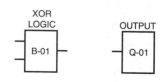

I-01	I-02	Q-01 (ON or OFF)
OPEN	OPEN	
OPEN	CLOSED	
CLOSED	CLOSED	
CLOSED	OPEN	

4. Connect the two input switches (I-01 and I-02) to the input of the first XOR logic block (B-01). Connect the output of the first logic block and switch three (I-03) to the input of the second XOR logic block (B-02) and the output of the second XOR logic block to the output (Q-01).

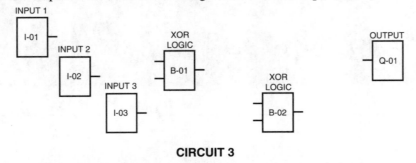

CIRCUIT 3

Programming Activity Objectives

- Using simulation software, program and test a three-input XOR logic circuit (Circuit 3).
- Verify the three-input XOR logic truth table based on observed operation of the circuit.

Procedure 19-2

Program switches connected to logic blocks to control an output, test the circuit using the TECO SG2 Client simulation software, and complete a truth table.

1. Open the TECO SG2 Client simulation software.

2. Open a new function block diagram document by clicking on the New FBD icon, located directly below the Help heading at the top of the screen.

3. Open a new circuit by clicking of the New Circuit icon, located directly below the File heading at the top of the screen.

4. In the Select Type window, select an SG2-12HR-D model.

5. Using the tool pallet located at the bottom of the screen, place three inputs (I-01, I-02, and I-03) on the left side of the screen. Inputs are found by opening the Co icon and clicking on I-IN for inputs.

6. Using the tool pallet located at the bottom of the screen, place one output (Q-01) on the right side of the screen. Outputs are found by opening the Co icon and clicking on Q-OUT for outputs.

7. Place two XOR logic blocks between the inputs and output. Logic blocks are found by opening the LB tool pallet, located at the bottom of the screen.

8. Using the Connect icon located at the bottom of the screen, connect the inputs to the left side of the logic block, and connect the output to the right side of the logic block. Connect the lines as required.

9. Select the Simulator icon, located to the left of the Run icon at the top of the screen.

10. Using the Input Status Tool window, test the circuit by clicking on each input (I1, I2, and I3) to turn each switch ON and OFF. Observe the circuit and status conditions of the inputs and output.

11. Complete the truth table.

12. End the simulation by clicking the Stop icon, located at the top of the screen.

13. Print a copy of the circuit by clicking the Print icon, located at the top of the screen.

14. Save the program as required.

INPUTS			OUTPUT
I-01	**I-02**	**I-03**	**Q-01 (ON or OFF)**
OPEN	OPEN	OPEN	
OPEN	OPEN	CLOSED	
CLOSED	CLOSED	OPEN	
OPEN	CLOSED	CLOSED	
CLOSED	CLOSED	CLOSED	

Name _____ **Date** _____

True-False

T   F   **1.** Normally, the applied voltage should be within ±10% of the motor's rated voltage.

T   F   **2.** Commercial power needs include the same needs as residential (1ϕ) and industrial (3ϕ) systems.

T   F   **3.** A 120/240 V, 3ϕ, 4-wire service is used to supply customers that require large amounts of 3ϕ power with some 120 V and 240 V, 1ϕ power.

T   F   **4.** Variation in applied voltage to a heating element has no effect on the amount of heat produced.

T   F   **5.** A footcandle is the unit used to measure the total amount of light produced by a light source.

T   F   **6.** The voltage on incandescent lamps can vary from 0 V to 10% higher than the lamp's rated voltage

T   F   **7.** A screw base is a bulb base that has two pins located on opposite sides.

T   F   **8.** Cool white or bright white fluorescent bulbs produce a pale blue-green whiteness.

Multiple Choice

_____ **1.** A ___ lamp is an incandescent lamp filled with a halogen gas.
   A. fluorescent
   B. low-pressure sodium
   C. tungsten-halogen
   D. mercury-vapor

_____ **2.** A(n) ___ circuit is a fluorescent lamp-starting circuit that has separate windings to provide continuous heating voltage on the lamp cathodes.
   A. instant-start
   B. instant-stop
   C. preheat
   D. rapid-start

201

_____ 3. ___ is the appearance of a color when illuminated by a light source.
    A. Color rendering
    B. Brightness
    C. Contrast
    D. Intensity

_____ 4. A ___ ballast is a ballast that uses two coils to regulate both voltage and current.
    A. reactor
    B. constant-wattage autotransformer
    C. high-reactance autotransformer
    D. two-winding, constant-wattage

_____ 5. The most common bulb shape is the "___" bulb.
    A. A
    B. P
    C. S
    D. T

Completion

_____ 1. A(n) ___ is an assemblage of equipment installed for switching, changing, or regulating the voltage of electricity.

_____ 2. A(n) ___ lamp is an electric lamp that produces light by the flow of current through a tungsten filament inside a gas-filled, sealed glass bulb.

_____ 3. A(n) ___ lamp is a low-pressure discharge lamp in which ionization of mercury vapor transforms ultraviolet energy generated by the discharge into light.

_____ 4. ___ is the ratio of brightness between different objects.

_____ 5. A(n) ___ is a conductor with a resistance high enough to cause the conductor to heat.

_____ 6. A(n) ___ is the light-producing element of an HID lamp.

_____ 7. A(n) ___ conductor is a conductor that does not normally carry current, except during a fault (short circuit).

_____ 8. A(n) ___ is an output device that converts electrical energy into light.

_____ 9. A(n) ___ is a preprinted peel-off marker designed to adhere when wrapped around a conductor.

_____ 10. A(n) ___ conductor is a current-carrying conductor that is intentionally grounded.

_____ 11. ___ is the effect that occurs when light falls on a surface.

_____ **12.** A(n) ___ conductor is a current-carrying conductor that is connected to loads through fuses, circuit breakers, and switches.

_____ **13.** ___ is the perceived amount of light reflecting from an object.

_____ **14.** ___ temperature is the temperature at which a lamp delivers its peak light output.

_____ **15.** A(n) ___ circuit is a fluorescent lamp-starting circuit that provides sufficient voltage to strike an arc instantly.

Lamp Identification

_____ **1.** Metal-halide

_____ **2.** Fluorescent

_____ **3.** Low-pressure sodium

_____ **4.** Incandescent

_____ **5.** Mercury-vapor

_____ **6.** Tungsten-halogen

_____ **7.** High-pressure sodium

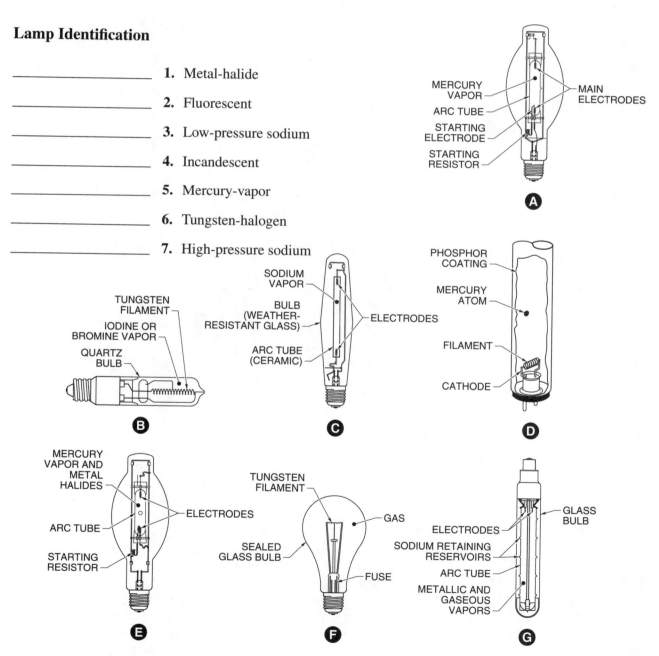

Name _____ Date _____

120/240 V, 1φ, 3-Wire Service Wiring

1. Connect each load to the service panel.

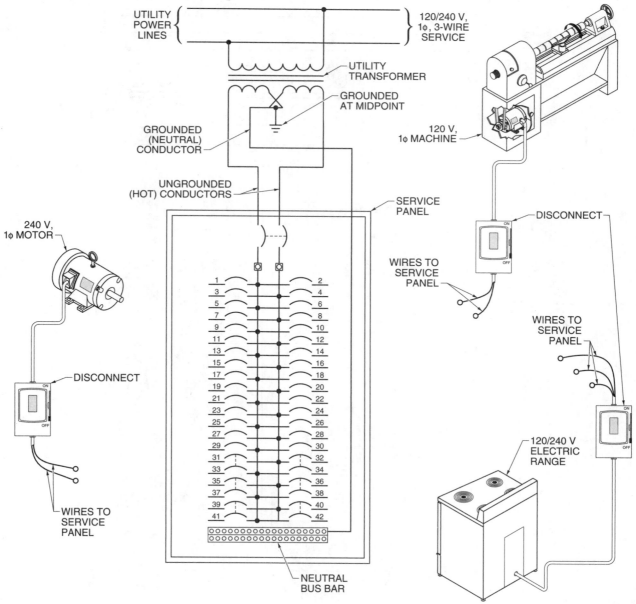

Name _____ **Date** _____

120/208 V, 3φ, 4-Wire Service Wiring

1. Connect each load to the service panel.

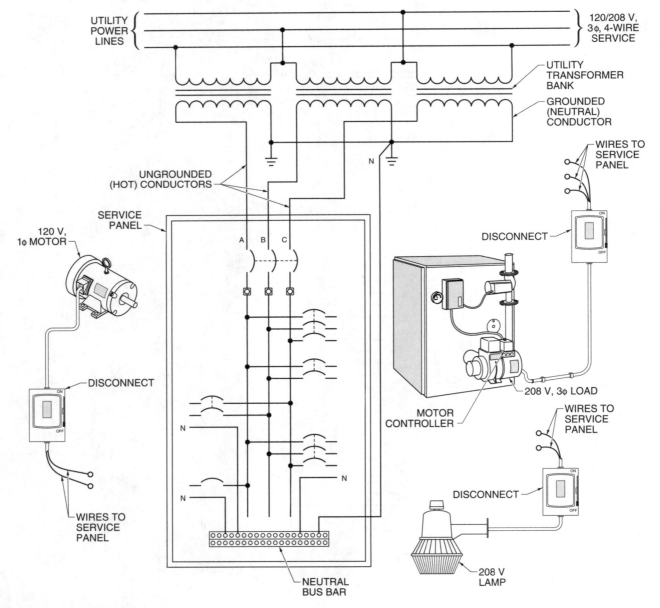

Name _____ Date _____

120/240 V, 3ϕ, 4-Wire Service Wiring

1. Connect each load to the service panel.

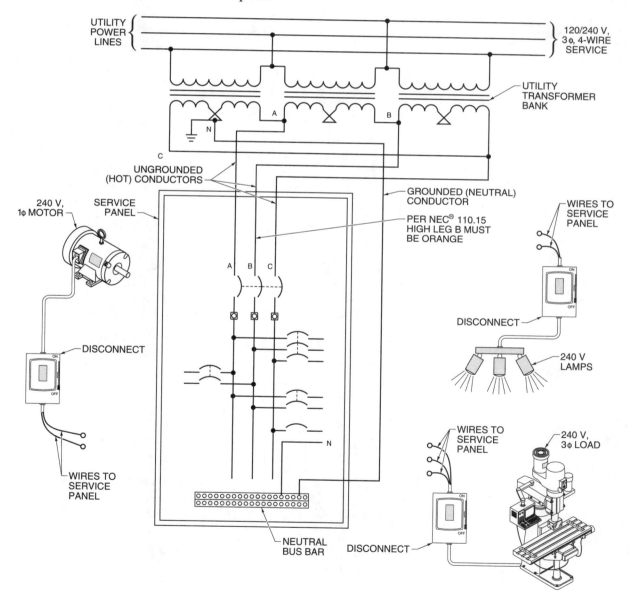

Name _____ **Date** _____

277/480 V, 3φ, 4-Wire Service Wiring

1. Connect each load to the service panel.

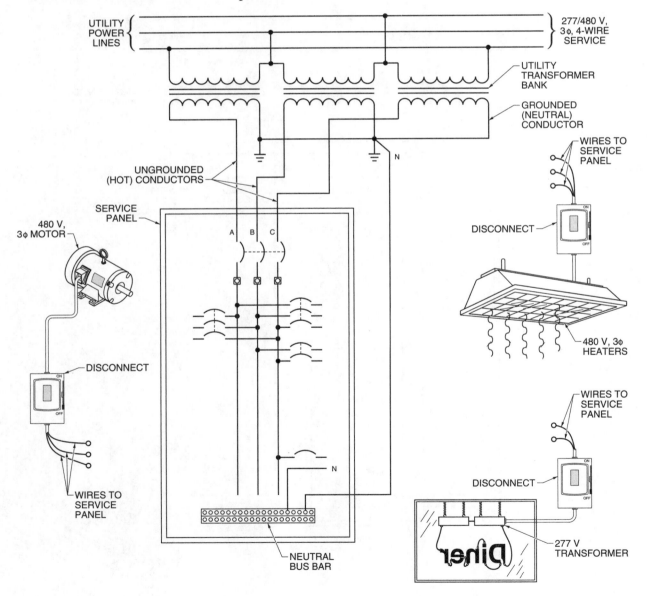

Name _____ Date _____

Conductor Color Coding

Determine each conductor color using standard conductor color coding.

_____ **1.** Conductor A color = ___

_____ **2.** Conductor B color = ___

_____ **3.** Conductor C color = ___

_____ **4.** Conductor D color = ___

_____ **5.** Conductor E color = ___

_____ **6.** Conductor F color = ___

_____ **7.** Conductor G color = ___

_____ **8.** Conductor H color = ___

_____ **9.** Conductor I color = ___

_____ **10.** Conductor J color = ___

_____ **11.** Conductor K color = ___

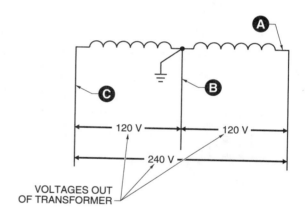

VOLTAGES OUT OF TRANSFORMER

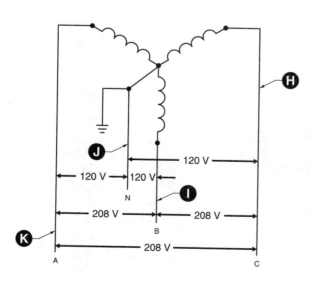

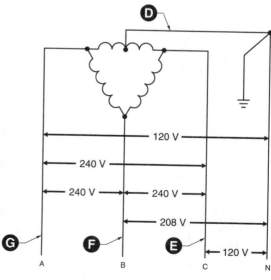

Name _____ Date _____

Lamp Wiring

Wire the ballast, capacitor, and lamp base to the power supply using the wiring diagram.

1.

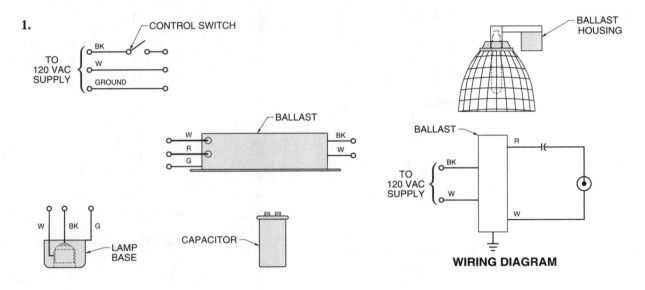

2.

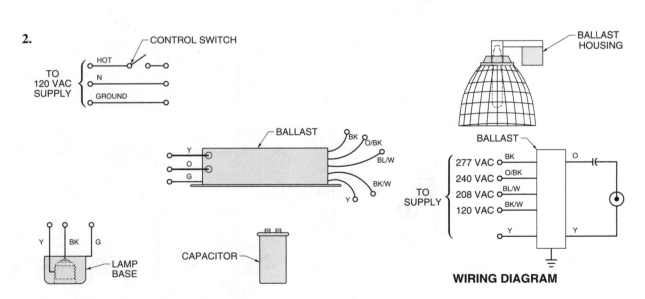

Name _____ **Date** _____

Objectives

- Use manufacturer specification data to determine the amount of light (in lumens) each bulb type produces.
- Use manufacturer specification data to determine the average rated lamp life for each bulb type.

Procedure 20-1

Lamps are used to produce visible light. The method used by the lamp bulb to produce light gives the bulb its name, such as incandescent, fluorescent, mercury-vapor, high-pressure sodium, etc. Each bulb type has advantages and disadvantages. Selecting the best bulb for an application requires an understanding of the advantages and disadvantages of each type. Click on the Lamp Data Sheets button on the CD-ROM. Next, click on the T5 Fluorescent Lamp button.

_____ **1.** The initial lumens output of a 28 W bulb is ___ lm.

_____ **2.** This value is equivalent to ___ lm/W.

_____ **3.** The rated life of the bulb is ___ hr.

Click on the Energy Saver Dimmable Lamp button.

_____ **4.** The initial lumens output of a 20 W bulb is ___ lm.

_____ **5.** This value is equivalent to ___ lm/W.

_____ **6.** The rated life of the bulb is ___ hr.

Click on the Compact Fluorescent Lamp button.

_____ **7.** The initial lumens output of a 26 W bulb is ___ lm.

_____ **8.** This value is equivalent to ___ lm/W.

_____ **9.** The rated life of the bulb is ___ hr.

Click on the Metal Halide Lamp button.

_____ **10.** The initial lumens output of a 50 W, 13691-0 bulb is ___ lm.

_____ **11.** This value is equivalent to ___ lm/W.

_____ **12.** The rated life of the bulb is ___ hr.

Click on the High-Pressure Sodium Lamp button.

_____ **13.** The initial lumens output of a 225 W bulb is ___ lm.

_____ **14.** This value is equivalent to ___ lm/W.

_____ **15.** The rated life of the bulb is ___ hr.

_____ **16.** Of all the bulb types in Procedure 20-1, the ___ bulb type produces the greatest output of light per watt.

_____ **17.** Of all the bulb types in Procedure 20-1, the ___ bulb type has the longest rated operating life.

Printreading Activity Objectives

- Determine the type of electrical services that can be used to deliver power to a given electrical system.

Procedure 20-2

Electrical services deliver power to electrical systems. The service selected must provide the correct voltage and current levels. The voltage provided should be within +5% to –10% of the voltage rating of the loads. The current provided should be greater (at least 10%) than the sum of all loads that might be energized at any one time. Click on the Compactor Print button on the CD-ROM and answer the following questions.

_____ **1.** Can the system be connected to a 120/240 V, 1φ, 3-wire service?

_____ **2.** Can the system be connected to a 120/208 V, 3φ, 4-wire service?

_____ **3.** Can the system be connected to a 120/240 V, 3φ, 4-wire service?

_____ **4.** Can the system be connected to a 277/480 V, 3φ, 4-wire service?

Name _____ **Date** _____

True-False

T   F   **1.** Industrial heating, ventilating, and air conditioning (HVAC) systems are basically the same as residential and commercial systems except for the size and number of units used.

T   F   **2.** Line diagrams are much harder to draw than pictorial drawings, power circuits, and wiring diagrams.

T   F   **3.** Any wire that is prewired when the component is purchased is normally not assigned a wire-reference number.

T   F   **4.** A line diagram is intended to show the physical relationship of the various devices in the control cabinet.

T   F   **5.** Electric heat is often used in applications that require heat for the production of goods because it is easy to control the temperature output of electric heating elements.

T   F   **6.** Most industrial circuits include a system for moving a product along a line and performing some function to the product.

T   F   **7.** Line numbers and numerical cross-reference numbers are used when reading the circuit's logic and operation.

T   F   **8.** A numerical cross-reference system allows the simplification of complex line diagrams.

T   F   **9.** A converter is an electronic device that changes DC voltage into AC voltage.

T   F   **10.** A digital signal is a type of input signal that can be either varying voltage or varying current.

213

Multiple Choice

_____ 1. Most industrial circuits, regardless of their size or type, have similarities to ___ circuits.
   A. other industrial
   B. commercial
   C. residential
   D. all of the above

_____ 2. A(n) ___ is a conductor that offers enough resistance to produce heat when connected to an electrical power supply.
   A. busway
   B. heating element
   C. electric motor drive
   D. conduit run

_____ 3. Steel mills and ___ are industrial applications that require simultaneous operation of two or more motors.
   A. paper mills
   B. bottling plants
   C. canning plants
   D. all of the above

_____ 4. Circuit design and modifications to existing circuits are possible using ___.
   A. bar graphs
   B. pie charts
   C. line diagrams
   D. scatterplots

_____ 5. Relays, contactors, and magnetic motor starters normally have more than ___ set(s) of auxiliary contacts.
   A. one
   B. two
   C. three
   D. four

Completion

_____ 1. ___ electrical circuits are circuits that are used to convert raw materials into finished products.

_____ 2. ___ diagrams use industrial electrical symbols to provide the information necessary to understand the operation of any electrical control circuit.

_____ 3. A(n) ___ is a solid-state programmable control module used to control an AC motor.

_____ 4. A(n) ___ is a solid-state control device that can be programmed to automatically control electrical systems in residential, commercial, and industrial facilities.

_____ 5. ___ are normally the largest part of industrial electrical circuits.

_____ 6. ___ is the physical connection of electrical components.

_____ 7. ___ is the frequent starting and stopping of a motor for short periods of time.

_____ 8. A(n) ___ is a metal-enclosed distribution system of busbars available in prefabricated sections.

_____ 9. ___ are used to identify and separate the different component parts (coil, NC contacts, etc.) included on the individual devices.

_____ 10. A(n) ___ diagram is a diagram which shows, with single lines and symbols, the logic of an electrical circuit or system of circuits and components.

_____ 11. A(n) ___ system consists of numbers in parenthesis to the right of the line diagram.

_____ 12. Wire-reference and ___ numbers are used when installing and troubleshooting the circuit.

_____ 13. A(n) ___ is an electronic device that controls the direction, speed, and torque of an electric motor in addition to providing motor protection and monitoring functions.

_____ 14. A(n) ___ switch performs two different switching functions, such as forward and reverse.

_____ 15. ___ is a control function that keeps a motor running after the start pushbutton is released.

Line Diagram Numbers

_____ 1. Manufacturer's terminal

_____ 2. Line

_____ 3. Numerical cross-reference

_____ 4. Wire reference

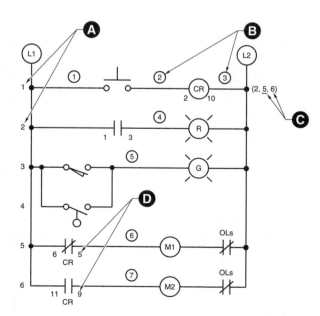

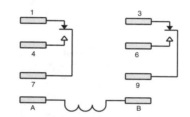

Industrial Circuits

Worksheet 21-1

21

Name _____ **Date** _____

Reference Number Addition

1. Add line number references, numerical cross-reference numbers, wire-reference numbers, and manufacturer's relay terminal numbers to the control circuit.

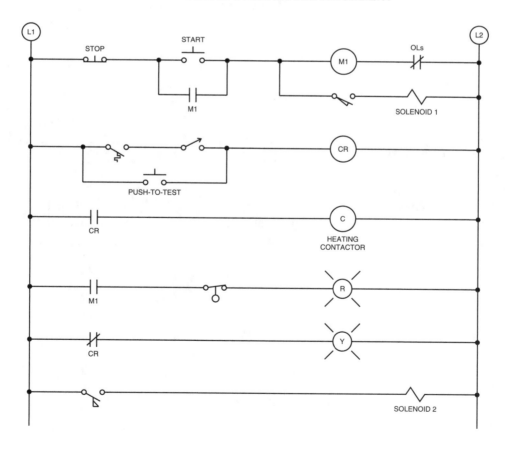

RELAY CONTACT ARRANGEMENT

Name _____ **Date** _____

Control Switch Addition

1. Redraw control circuit 1 adding three stop pushbuttons and three start pushbuttons. Include line number references, numerical cross-reference numbers, and wire-reference numbers.

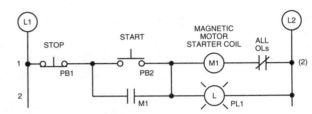

CONTROL CIRCUIT 1

2. Redraw control circuit 2 adding a temperature switch that automatically stops both motors if a set temperature is reached, a level switch that does not allow motor 1 (M1) to start unless the level in the tank is at a set level, and a pressure switch that automatically stops motor 2 (M2) if the pressure in the system reaches a set pressure. Include line number references, numerical cross-reference numbers, and wire-reference numbers.

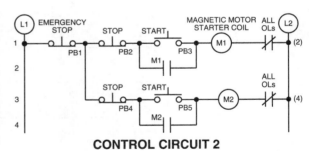

CONTROL CIRCUIT 2

Name _____ Date _____

Sequence Control Circuit

1. Connect three motor starters (M1, M2, and M3) so that M3 cannot start unless M2 is energized and M2 cannot start unless M1 is energized. Each motor starter is controlled from its own standard start/stop pushbutton station. Include line number references, numerical cross-reference numbers, and wire-reference numbers.

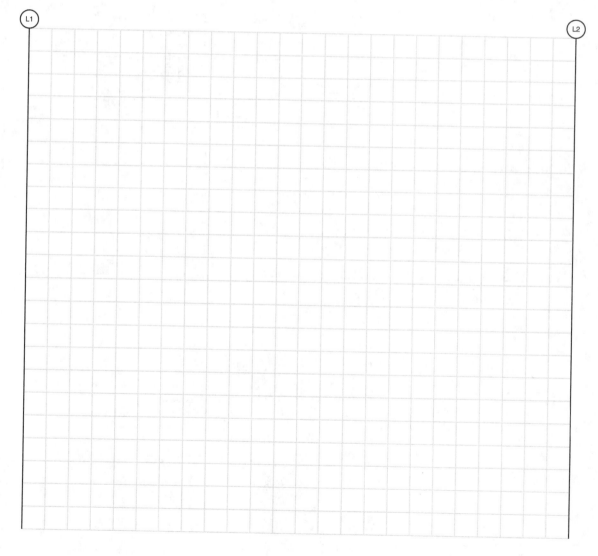

Name _____ **Date** _____

Indicator Lamp Addition

1. Add a red indicator lamp to the control circuit to indicate when the selector switch controlling the heating element is in the ON position and a green indicator lamp to indicate when the selector switch controlling the heating element is in the OFF position.

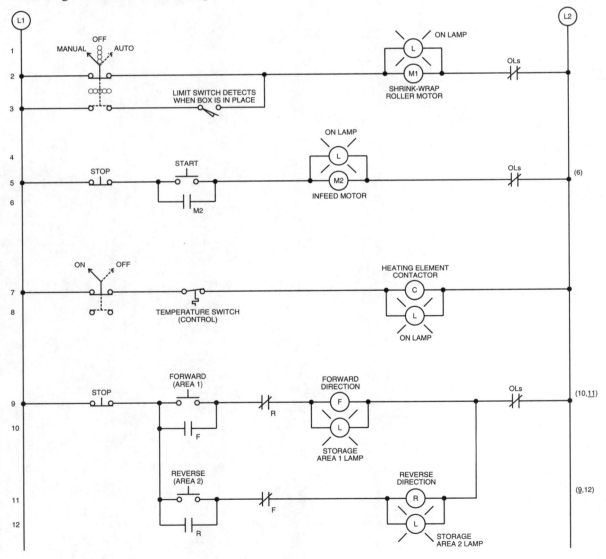

Name _____ **Date** _____

Interlocking Circuits

1. Redraw the control circuit so that motor 1 (M1) cannot be ON at the same time as motor 2 (M2). *Note:* Assume that each pushbutton and/or starter has extra contacts. Include line number references, numerical cross-reference numbers, and wire-reference numbers.

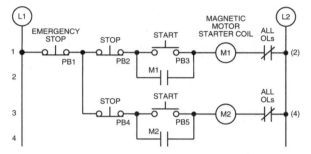

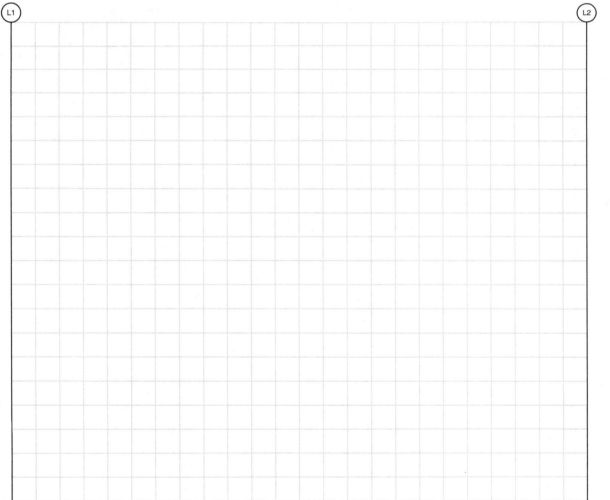

Name _____ Date _____

Manual Control Switch Addition

1. Redraw the reversing circuit adding a second reverse pushbutton, a second forward pushbutton, and a second stop pushbutton. Include line number references, numerical cross-reference numbers, and wire-reference numbers.

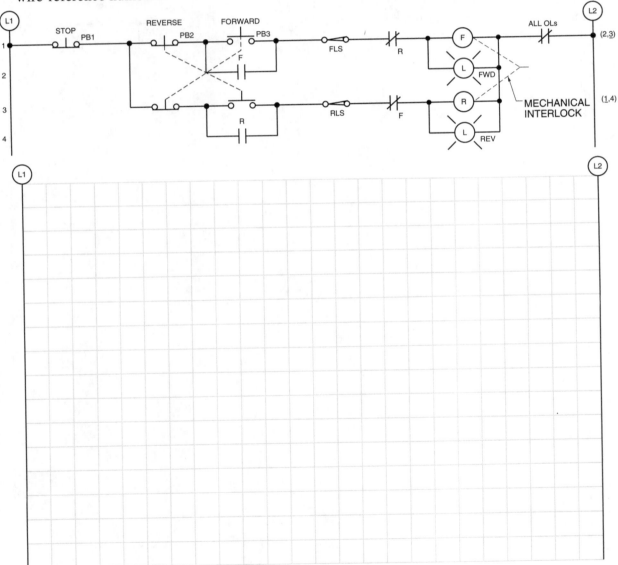

Name _____ Date _____

Hands-On Activity Objectives

- Build a circuit that shows how a switch can be used as a dual-function switch.
- Build a circuit that shows how a switch can be used as a single-function switch.
- Build a circuit that uses a relay as an interface to allow a small switching current to control a high load current.

Procedure 21-1

All switches are either dual-function or single-function switches. A dual-function switch performs two different switching functions, such as turning ON and turning OFF. A single-function switch performs only one switching function, such as only turning ON or only turning OFF. Single-function switches are used in industrial motor control circuits as start, stop, jog, forward, reverse, etc., pushbuttons. The normally open contacts on relays, contactors, and motor starters are used to develop circuit memory that allows a switch to operate as a single-function switch.

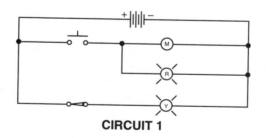

CIRCUIT 1

Build and test Circuit 1.

_____ 1. Does each switch perform a dual switching function of turning ON the load and turning OFF the load?

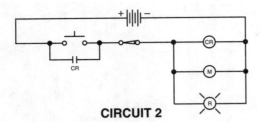

CIRCUIT 2

Build and test Circuit 2.

_____ **2.** Does each switch perform only a single switching function of turning ON the load or turning OFF the load?

In Circuit 2, the relay is used to develop circuit memory so that the ON switch can be pressed and released without turning the loads OFF. Relays are also used as interfaces to allow a small switching current to control high-power loads.

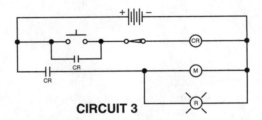

CIRCUIT 3

Build and test Circuit 3.

_____ **3.** Does the pushbutton carry the current of the loads?

Measurement Activity Objectives

• Measure the current in a circuit that uses a relay as an interface to verify that by using a relay the switch does not carry the high current of the loads.

Procedure 21-2

Set a DMM to measure in-line DC current. Open Circuit 3 between the pushbutton and relay coil and insert the DMM.

_____ **1.** With the loads ON, the amount of current flowing through the pushbutton is ___ mA.

Remove the DMM and reconnect the circuit. Open Circuit 3 between the relay's normally closed contact and the loads and insert the DMM.

_____ **2.** With the loads ON, the amount of current flowing through the relay contact is ___ mA.

Printreading Activity Objectives

- Use an electrical print to determine the placement of test leads when taking voltage measurements and troubleshooting an electrical circuit.
- Use an electrical print to identify replacement-part electrical requirements.

Procedure 21-3

Click on the Compactor Print button on the CD-ROM and answer the following questions. When troubleshooting inside a control enclosure, a DMM set to measure voltage can be used to take voltage measurements.

_____ **1.** On which two terminals (at the terminal strip marked X1 through 29) would the DMM leads be connected to verify that the 120 VAC control circuit is powered and fuse 1FU is good?

Control relay 1CR needs to be replaced.

_____ **2.** What is the voltage requirement of the relay coil?

_____ **3.** What number of normally open contacts is required?

_____ **4.** What number of normally closed contacts is required?

In the control circuit, it is thought that the amber light is burned out. A fused jumper has been placed across terminals 6 and 28 (pressure switch P/S1).

_____ **5.** On which two terminals would the DMM leads be connected to verify that the amber lamp has voltage, and thus is burned out and in need of replacing?

Fluid Power Circuits

Review Questions

22

Name _____ **Date** _____

True-False

T F **1.** A check valve is a valve that allows the fluid to flow in one direction only.

T F **2.** A position is a flow path through a valve.

T F **3.** An actuator is a device that converts fluid energy into mechanical motion.

T F **4.** A valve is a device that converts fluid energy into a linear mechanical force.

T F **5.** A normally closed valve is a valve that does not allow pressurized fluid to flow out of the valve in the spring-actuated position.

Multiple Choice

_____ **1.** A(n) ___ motor is a motor that converts pressurized fluid force into rotating motion.
   A. pressure relief
   B. fluid power
   C. actuator
   D. safety

_____ **2.** A ___ is the number of positions within a valve in which the spool is placed to direct fluid through the valve.
   A. directional control valve
   B. position
   C. way
   D. rotary actuator

_____ **3.** A ___ moves back and forth over a fixed arc that is less than one complete revolution (360 degrees).
   A. directional control valve
   B. rotary actuator
   C. pushbutton
   D. lever

© 2013 American Technical Publishers, Inc.
All rights reserved

_____ **4.** A ___ requires pressurized fluid in only one port.
   A. two-way valve
   B. hydraulic cylinder
   C. directional control valve
   D. single-acting cylinder

_____ **5.** A ___ valve is a valve that connects, disconnects, or directs the fluid pressure from one part of a circuit to another.
   A. normally closed
   B. directional control
   C. normally open
   D. two-way

Completion

_____ **1.** ___ is the transmission and control of energy by means of a pressurized fluid.

_____ **2.** A(n) ___ is a device that increases the pressure of atmospheric air to the desired pressure of the receiver tank.

_____ **3.** ___ is the transmission of energy using a liquid (normally oil).

_____ **4.** A(n) ___ valve is a valve used to control the volume of fluid that flows in a part of the circuit.

_____ **5.** A(n) ___ is a liquid or a gas that can move and change shape without separating when under pressure.

_____ **6.** A(n) ___ pump is a pump that delivers a definite quantity of fluid for every stroke, revolution, or cycle.

_____ **7.** ___ is the transmission of energy using air.

_____ **8.** A(n) ___ cylinder is a cylinder that requires fluid pressure to advance and retract the cylinder.

_____ **9.** A(n) ___ is a fluid substance that can flow readily and assume the shape of its container.

_____ **10.** A(n) ___ valve is a valve that limits the amount of pressure in a fluid power system.

Fluid Power Symbols

_____ **1.** Four-way, 2-position

_____ **2.** Solenoid

_____ **3.** Lever

_____ **4.** Flow control with check valve

_____ **5.** Two-way, 2-position

Name _____ Date _____

Product Flattening Application

1. A roller is used to flatten a moving product with pressure from a hydraulic cylinder. Connect the hydraulic cylinder to a 3-position, double solenoid-operated, spring-centered valve. The center position of the valve must block all ports, allowing the cylinder to be held in a fixed position. Include a flow control with check valve to control the speed of the extend stroke of the cylinder. Connect the outlet of the hydraulic pumping unit to the inlet of the valve. Draw the electrical circuit so that a pushbutton controls the extend solenoid and either one of two different pushbuttons control the retract solenoid.

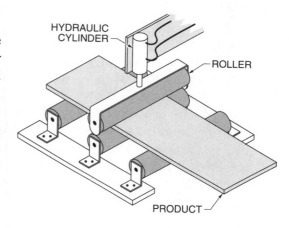

HYDRAULIC CYLINDER — ROLLER — PRODUCT

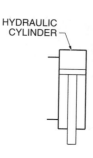

HYDRAULIC CYLINDER

L1   L2

HYDRAULIC PUMPING UNIT OUTLET

RELIEF VALVE

HYDRAULIC PUMP — FILTER — ELECTRIC MOTOR — M

HYDRAULIC PUMPING UNIT — RESERVOIR

Name _____   Date _____

Part Eject Application

1. A part that is upside down is sensed by a proximity switch and ejected by a blast of air from an ejector valve. Connect the outlet of the FRL to a 2-position, solenoid-operated, spring-return (ejector) valve used to supply the air that ejects the part. Include a flow control valve used to set the amount of air pressure to the ejector valve. Draw the electrical circuit so that proximity switch turns ON a timer that is used to energize the ejector valve solenoid for 1 sec when it detects an upside-down part. Include a test pushbutton that is used at any time to test the ejector valve.

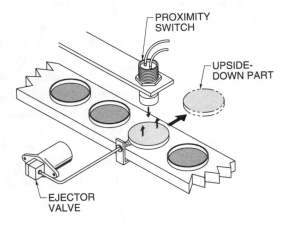

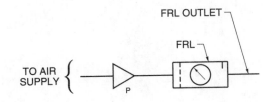

Name _____ **Date** _____

Backhoe Application

1. Three double-acting cylinders are used to control the movement of the bucket of a backhoe. Connect each cylinder
to its own maintained (detent) 3-position, lever-operated,
directional control valve. The center position of each valve
must block all ports, allowing each cylinder to be held in a
fixed position. Include two flow control with check valves
for each cylinder to control the amount of fluid (speed) flowing into each cylinder. Connect the outlet of the hydraulic
pumping unit to the inlet of the valves.

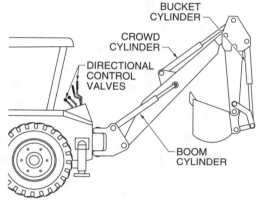

BUCKET
CYLINDER

CROWD
CYLINDER

DIRECTIONAL
CONTROL
VALVES

BOOM
CYLINDER

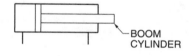 BOOM
CYLINDER

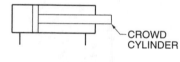 CROWD
CYLINDER

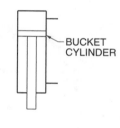 BUCKET
CYLINDER

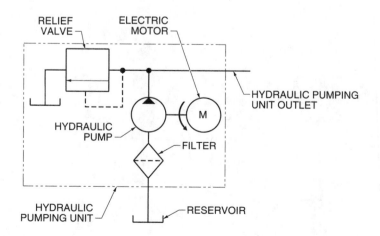

RELIEF
VALVE

ELECTRIC
MOTOR

HYDRAULIC PUMPING
UNIT OUTLET

HYDRAULIC
PUMP

M

FILTER

HYDRAULIC
PUMPING UNIT

RESERVOIR

Name _____ Date _____

Part Feed Application

1. Two pneumatic cylinders are used to hold and release parts fed into a production line. The cylinders alternate so that as one advances, the other retracts. Connect each cylinder to its own 2-position, solenoid-operated, spring return valve. Connect the outlet of the FRL to the inlet of the valves. Draw the electrical circuit so that when the recycle timer is energized, the hold solenoid energizes (but not the release solenoid) and after timing out, the release solenoid energizes (and the hold solenoid de-energizes). The recycle time keeps the two solenoids alternating ON and OFF as long as the timer is energized.

HOLD CYLINDER

RELEASE CYLINDER

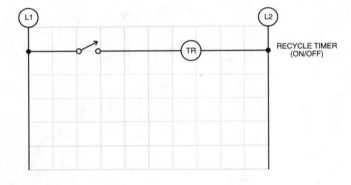

RECYCLE TIMER (ON/OFF)

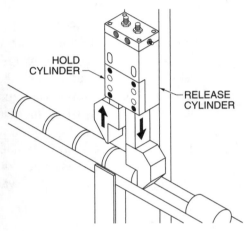

HOLD CYLINDER

RELEASE CYLINDER

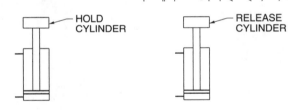

FRL OUTLET

FRL

TO AIR SUPPLY

P

Name _____ Date _____

Pick-and-Place Robot Application

1. Double-acting cylinders are used to control the movement of a pneumatic robotic arm. Connect each cylinder to its own 2-position, double solenoid-operated valve. Add a flow control valve to each cylinder to control the speed of the cylinder when extending and retracting. Connect the outlet of the FRL to the inlet of the valves. Draw the electrical circuit so the PB1 activates the solenoid controlling the vertical cylinder extend stroke, PB2 activates the solenoid controlling the vertical cylinder retract stroke, PB3 activates the solenoid controlling the horizontal cylinder extend stroke, PB4 activates the solenoid controlling the horizontal cylinder retract stroke, PB5 activates the solenoid controlling the gripper cylinders open movement, and PB6 activates the solenoid controlling the gripper cylinders close movement.

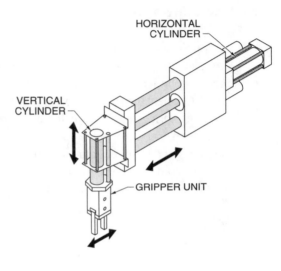

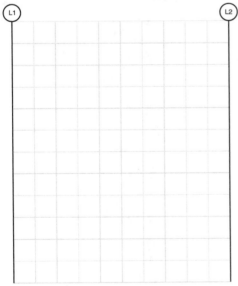

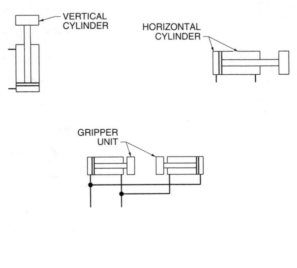

Name _____ Date _____

Part Rotating Application

1. A double-acting pneumatic cylinder and rotary actuator are used to rotate a part on a production line. Connect the lift cylinder to a 2-position, double solenoid-operated valve. Connect the rotary actuator to a 2-position, solenoid-operated, spring-return valve. Include two flow control with check valves to control the speed of the lift cylinder when extending (lift) and retracting (drop). Include a flow control valve to control the speed of the rotary actuator. Connect the outlet of the FRL to the inlet of the valves.

LIFT
CYLINDER

ROTARY
ACTUATOR

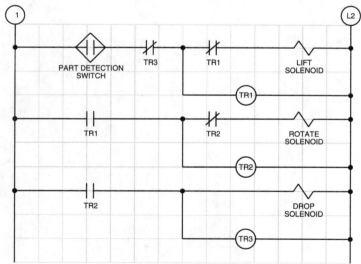

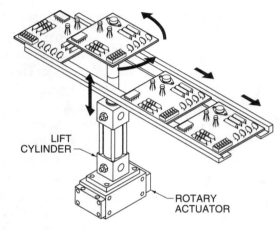

LIFT
CYLINDER

ROTARY
ACTUATOR

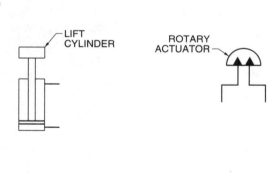

FRL OUTLET

FRL

TO AIR
SUPPLY

P

Name _____ **Date** _____

Programming Activity Objectives

- Program a glue gun application circuit.
- Add documentation to the circuit and test the circuit using simulation software.
- Program a conveyor positioning circuit.
- Add documentation to the circuit and test the circuit using simulation software.

Procedure 22-1

1. Open the TECO SG2 Client simulation software.

2. Program the Glue Gun circuit.

3. Add all documentation to the circuit.

4. Test the circuit and reprogram any required changes to make the circuit operate as intended.

5. Print a copy of the circuit.

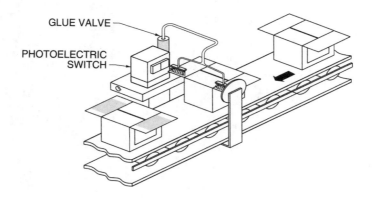

GLUE VALVE

PHOTOELECTRIC
SWITCH

Glue Gun Application

A single solenoid-operated valve is used to control a glue gun in the glue gun circuit. In this circuit, timer 1 (T1) is set for a 1.5 sec delay to delay the glue spray so that no glue is sprayed on the edge of a box or on the conveyor belt. Timer 2 (T2) sprays the glue for 4.8 sec. The circuit inputs and outputs are used as follows:

Input 1 = glue gun ON/OFF selector switch

Input 2 = photoelectric switch that detects a box

Input 3 = push-to-test button to test the glue gun

Output 1 = "box in place" lamp

Output 2 = "glue gun OFF" lamp

Output 3 = glue gun

Output 4 = "glue gun ON" lamp

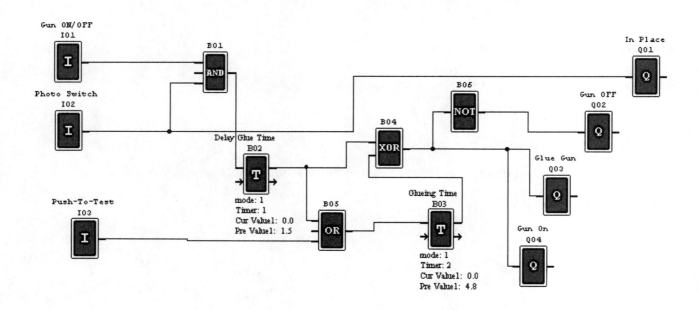

Procedure 22-2

1. Open the TECO SG2 Client software.

2. Program the Glue Gun circuit.

3. Add all documentation to the circuit.

4. Test the circuit and reprogram any required changes to make the circuit operate as intended.

5. Print a copy of the circuit.

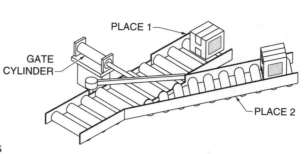

Conveyor Positioning Circuit

A double solenoid-operated valve is used to send boxes coming down a conveyor to place 1 or place 2 in the warehouse. An alarm is sounded for 5 sec before the gate changes position, to warn of the movement of the gate. The circuit input and outputs are used as follows:

Input 1 = conveyor-positioning selector switch

Output 1 = alarm to warn that the gate will be moving

Output 2 = solenoid that moves gate to move boxes to place 1

Output 3 = solenoid that moves gate to move boxes to place 2

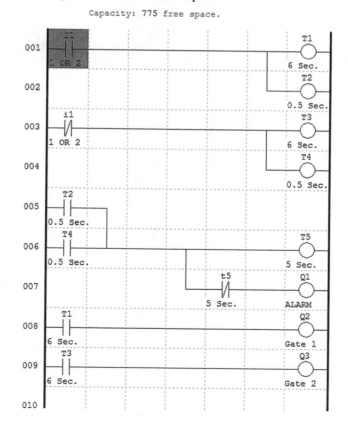

Audio Systems

Review Questions

23

Name _____ Date _____

True-False

T   F   **1.** Rarefaction is an area of increased pressure in a sound wave produced when a vibrating object moves outward.

T   F   **2.** Sound intensity is a measure of the amount of energy flowing in a sound wave.

T   F   **3.** A line matching transformer is an adapter containing a transformer that changes the three-wire balanced system into a two-wire unbalanced system.

T   F   **4.** The bel is the international unit of frequency and is equal to one cycle per second (cps).

T   F   **5.** Frequency is the time required to produce one complete cycle of a sound wave.

Multiple Choice

_____ **1.** The ___ value is a mathematical expression equal to 0.707 times the peak value of a waveform.
- A. peak
- B. peak-to-peak
- C. root-mean-square (rms)
- D. average

_____ **2.** A ___ system is a system that integrates a high-quality video source with multichannel audio electronics.
- A. home theater
- B. public address
- C. shelf
- D. portable

_____ **3.** ___ is the range of frequencies that a device can accept within tolerable limits.
- A. Compression
- B. Impedance
- C. Pitch
- D. Bandwidth

_____ **4.** A(n) ___ is a record player without an amplifier.
- A. compact disk player
- B. woofer
- C. receiver
- D. turntable

_____ **5.** A ___ is a small speaker that is designed to reproduce high frequencies between 4 kHz and 22 kHz.
- A. midrange speaker
- B. tweeter
- C. woofer
- D. peak-level speaker

_____ **6.** ___ is a raspy sound caused by a damaged component, such as a torn cone.
- A. Speaker noise
- B. Transient response
- C. Harmonic distortion
- D. Rarefaction

_____ **7.** ___ level is the signal level produced by instruments such as guitars and old keyboards.
- A. Microphone
- B. Line
- C. Instrument
- D. Audio

_____ **8.** A(n) ___ is a device that absorbs electrical interference from meters, etc. before these signals can travel down the power line.
- A. line matching transformer
- B. AC line filter
- C. phase filter
- D. transient filter

Completion

_____ **1.** ___ is energy that consists of pressure vibrations in the air.

_____ **2.** A(n) ___ is an electrical representation of a sound in the form of fluctuating voltage or current.

_____ **3.** A(n) ___ is an overtone whose frequency is a multiple of the fundamental frequency.

_____ **4.** A(n) ___ is an electric device that converts electrical signals into sound waves.

_____ **5.** A(n) ___ is an audio device that provides a signal to another audio device.

_____ **6.** ___ is the time relationship of a sine (or sound) wave to a known time reference.

_____ 7. ___ is a sharp clicking that occurs when a speaker cone cannot move as far as required by the audio signal.

_____ 8. ___ is an area of reduced pressure in a sound wave produced when a vibrating object moves inward.

_____ 9. ___ is any unintentional change in the characteristics of an audio signal.

_____ 10. A(n) ___ is an electric device that turns medium-level audio signals into strong audio signals that are sent to speakers.

_____ 11. A(n) ___ is an arrangement of electronic components that is designed to amplify sound.

_____ 12. ___ is the distance covered by one complete cycle of a given frequency sound as it passes through the air.

_____ 13. ___ is the highness or lowness of a sound.

_____ 14. The ___ spectrum is the part of the frequency spectrum that humans can hear (20 Hz – 20 kHz).

_____ 15. A(n) ___ is a tuner and amplifier contained in the same enclosure.

_____ 16. The ___ is the unit used to measure the intensity (volume) of sound.

_____ 17. A(n) ___ system is an audio system that uses two independent channels that are routed to a pair of speakers situated to the right and left of the listener.

_____ 18. The ___ spectrum is the range of all possible frequencies.

_____ 19. ___ is the number of air pressure fluctuation cycles produced per second.

_____ 20. ___ is the concept of expanding reproduced sound characteristics from one dimension (mono) to three dimensions.

Electrodynamic Speakers

_____ 1. Dust cap

_____ 2. Permanent magnet

_____ 3. Iron core

_____ 4. Spider

_____ 5. Frame

_____ 6. Voice coil

_____ 7. Cone

_____ 8. Suspension

Name _____ Date _____

Basic Speaker Connections

1. Connect the speakers to the amplifier so the total impedance of the speakers equals the impedance output rating of the amplifier.

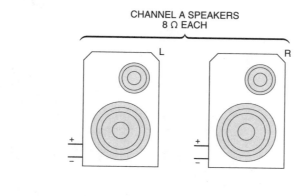

CHANNEL A SPEAKERS
8 Ω EACH

L   R

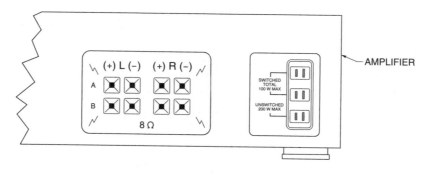

(+) L (−)   (+) R (−)

A

B

8 Ω

AMPLIFIER

SWITCHED
TOTAL
100 W MAX

UNSWITCHED
200 W MAX

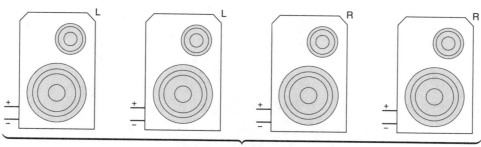

L   L   R   R

CHANNEL B SPEAKERS
4 Ω EACH

243

Name _____ Date _____

Combination Speaker Connections

Connect the speakers to the amplifier so the total impedance of the speakers equals the designated imped-ance output rating of the amplifier.

1.

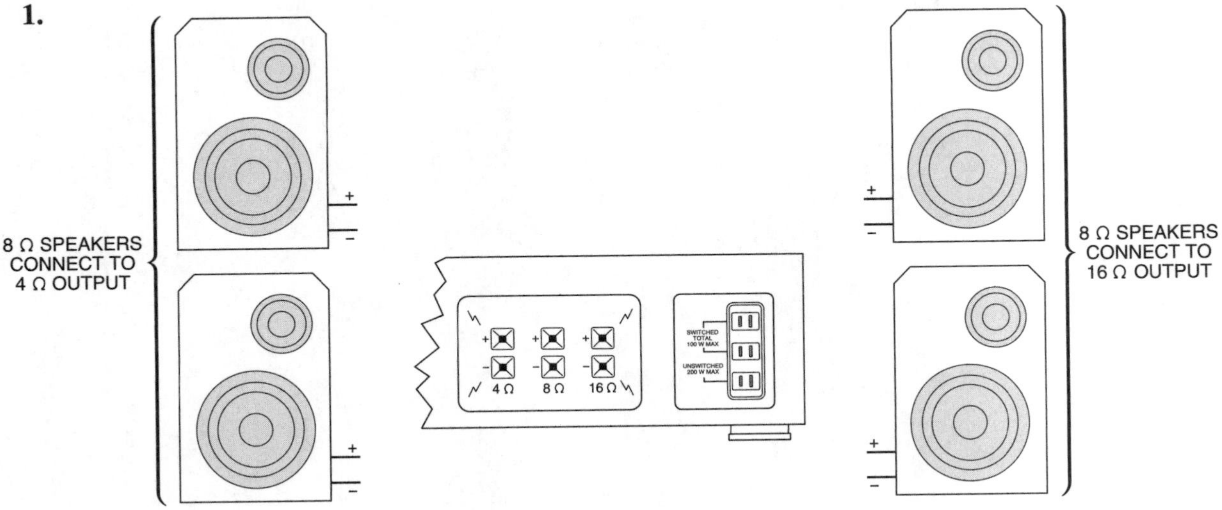

8 Ω SPEAKERS CONNECT TO 4 Ω OUTPUT

8 Ω SPEAKERS CONNECT TO 16 Ω OUTPUT

2.

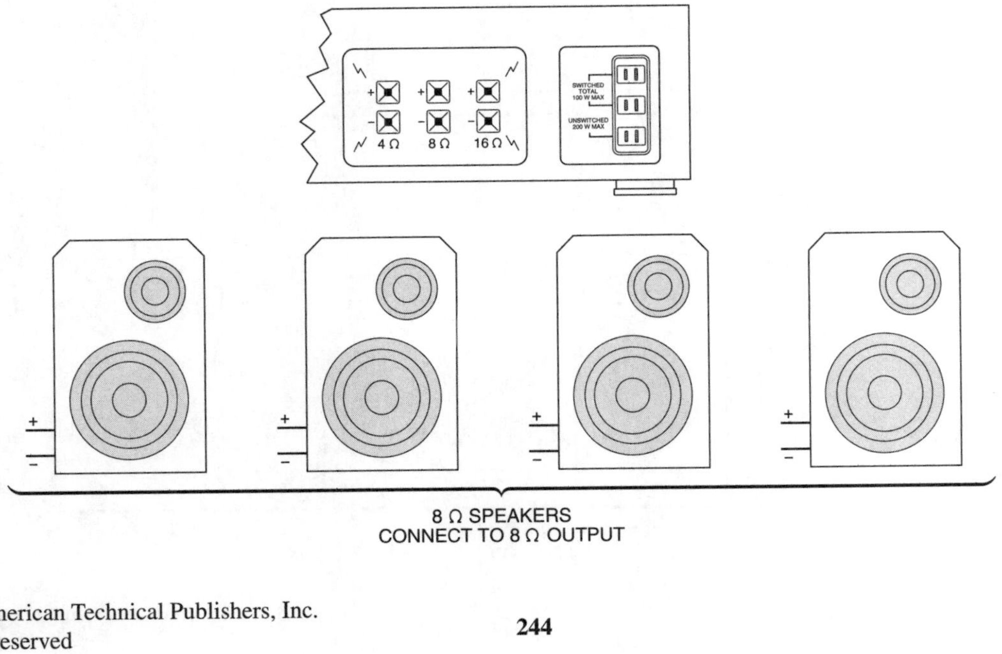

8 Ω SPEAKERS CONNECT TO 8 Ω OUTPUT

Name _____ **Date** _____

Speaker Wire Sizing

Determine the correct wire size for each speaker run.

_____ **1.** Wire size required at A = No. ___

_____ **2.** Wire size required at B = No. ___

RECOMMENDED SPEAKER WIRE SIZES*						
Amplifier Maximum Wattage Output**	4 Ω Speaker		8 Ω Speaker		16 Ω Speaker	
	1′ – 100′	101′ – 200′	1′ – 100′	101′ – 200′	1′ – 100′	101′ – 200′
25	14	12	16	14	16	14
50	12	10	14	12	16	14
100	10	***	12	10	14	12

* standard pair of speaker (in AWG No.)
** in W
*** not recommended

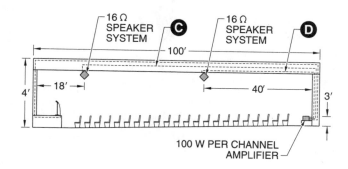

_____ **3.** Wire size required at C = No. ___

_____ **4.** Wire size required at D = No. ___

Name _____ Date _____

Basic Meter Testing

Connect each meter to the outlet and show dial setting to check for incorrect wiring to prevent electrical shock and excessive hum and noise in audio systems.

1. Connect meter 1 to measure the voltage from hot to neutral.

2. Connect meter 2 to measure the voltage from hot to ground.

3. Connect meter 3 to measure the voltage from neutral to ground.

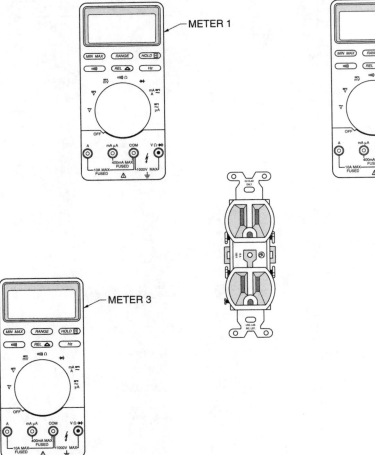

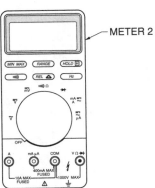

Audio Systems

Activities

23

Name _____ **Date** _____

Audio System Activity Objectives

- Listen to different frequencies within the human hearing range to observe the difference between low, middle, and high frequencies.
- Determine lower and upper frequency hearing limits.

Procedure 23-1

The ideal frequency range of human hearing is 20 Hz to 20 kHz. The hearing frequency range of an individual decreases as the person ages and/or the person's hearing is damaged. Click on the Audio Frequencies Interactive Simulation button on the CD-ROM. Listen to the tone at each of the frequencies listed and mark as can or cannot be heard.

_____ **1.** The 20 Hz frequency ___ be heard.

_____ **2.** The 50 Hz frequency ___ be heard.

_____ **3.** The 100 Hz frequency ___ be heard.

_____ **4.** The 250 Hz frequency ___ be heard.

_____ **5.** The 500 Hz frequency ___ be heard.

_____ **6.** The 1 kHz frequency ___ be heard.

_____ **7.** The 3 kHz frequency ___ be heard.

_____ **8.** The 5 kHz frequency ___ be heard.

_____ **9.** The 10 kHz frequency ___ be heard.

_____ **10.** The 12 kHz frequency ___ be heard.

_____ **11.** The 14 kHz frequency ___ be heard.

_____ **12.** The 15 kHz frequency ___ be heard.

_____ **13.** The 16 kHz frequency ___ be heard.

_____ **14.** The 17 kHz frequency ___ be heard.

_____ **15.** The 18 kHz frequency ___ be heard.

Hands-On Activity Objectives

- Build a circuit in which a pushbutton sounds a buzzer (or flashes a light) that can be used to send messages using Morse code.
- Transmit and interpret words using Morse code.

Procedure 23-2

The first electrical system for communication over distances was the telegraph system, which used electrical signals made up of dots, dashes, and spaces to send messages. Telegraph messages and news were transmitted using Morse code. Morse code is named after Samuel Morse, who patented the telegraph in 1840. This code is considered the simplest communication code used by military and civilian emergency personnel.

International Morse code

A	• —	N	— •	1	• — — — —
B	— • • •	O	— — —	2	• • — — —
C	— • — •	P	• — — •	3	• • • — —
D	— • •	Q	— — • —	4	• • • • —
E	•	R	• — •	5	• • • • •
F	• • — •	S	• • •	6	— • • • •
G	— — •	T	—	7	— — • • •
H	• • • •	U	• • —	8	— — — • •
I	• •	V	• • • —	9	— — — — •
J	• — — —	W	• — —	0	— — — — —
K	— • —	X	— • • —		
L	• — • •	Y	— • — —		
M	— —	Z	— — • •		

1. Build the following circuit.

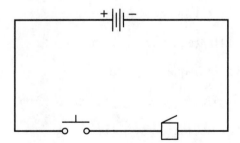

2. Write down your initials in Morse code.

3. Using the circuit built in Step 1, practice sending your initials in Morse code by pressing the pushbutton for long (dash) or short (dot) periods.

4. Send a word to another student in the class and have that student decode the word.

5. Have another student send a word and decode the word from the student.

Electronic Control Devices

Review Questions

Name _____ **Date** _____

True-False

| T | F | **1.** | Doping is the process by which the crystal structure of an atom is altered. |

T   F   **1.** Doping is the process by which the crystal structure of an atom is altered.

T   F   **2.** A silicon controlled rectifier is a three-terminal semiconductor thyristor that is normally an open circuit until a signal applied to the gate switches it to the conducting state in one direction.

T   F   **3.** Protons are particles with a positive electrical charge of one unit.

T   F   **4.** Pads are conducting paths used to connect components on a PC board.

Multiple Choice

_____ **1.** ___ are negatively charged particles whirling around the nucleus at great speeds in shells.
   A. Electrons
   B. Protons
   C. Neutrons
   D. Diacs

_____ **2.** A ___ is a three-terminal semiconductor thyristor that is triggered into conduction in either direction by a small amount of current to its gate.
   A. transistor
   B. diac
   C. triac
   D. UJT

_____ **3.** A(n) ___ is a three-terminal device that controls current through the device depending on the amount of voltage applied to the base.
   A. diac
   B. SCR
   C. diode
   D. transistor

_____ **4.** ___ are in the outermost shell of an atom.
   A. Protons
   B. Neutrons
   C. Valence electrons
   D. N-type materials

_____ **5.** A(n) ___ is the missing electrons in the crystal structure of a semiconductor.
   A. bus
   B. hole
   C. way
   D. edge card

Completion

_____ **1.** A(n) ___ is the smallest particle that an element can be reduced to and still maintain the properties of that element.

_____ **2.** A(n) ___ is a transistor consisting of N-type material with a region of P-type material doped within the N-type material.

_____ **3.** ___ are small round conductors to which component leads are soldered.

_____ **4.** A(n) ___ is an electronic component that allows current to pass through in only one direction.

_____ **5.** The ___ is the heavy, dense center of an atom which contains protons and neutrons and has a positive electrical charge.

_____ **6.** ___ devices are devices in which electrical conductivity is between that of a conductor and that of an insulator.

_____ **7.** A(n) ___ is a PC board with multiple terminations (terminal contacts) on one end.

_____ **8.** A(n) ___ is a metal extension from the transistor case.

_____ **9.** ___ are particles with no electrical charge.

_____ **10.** A(n) ___ is an insulating material such as fiberglass or phenolic with conducting paths laminated to one or both sides of the board.

_____ **11.** ___ electrons are electrons in the outermost shell of an atom.

_____ **12.** ___ material is material with empty spaces (holes) in its crystalline structure.

Electronic Symbols

_____ **1.** UJT

_____ **2.** Zener diode

_____ **3.** Triac

Ⓐ   Ⓑ   Ⓒ

Name _____ Date _____

Diode Current Flow

Diodes are used to allow current flow in only one direction. Normal current flow is from positive to negative. Diodes are connected to convert AC to DC. Draw an arrow to indicate the direction of current flow through each part of the circuit.

1.

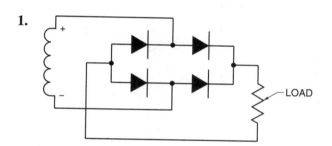

2.

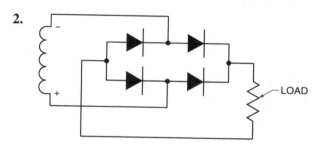

3.

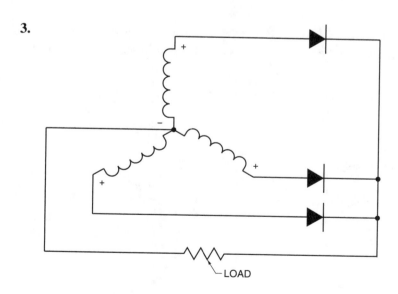

Electronic Control Devices

Worksheet 24-2

Name _____ **Date** _____

NPN Transistor Connections

1. Connect the NPN transistor in the circuit so that switch 1 and switch 2 controls the current flow to the transistor and the transistor controls the current flow to the load.

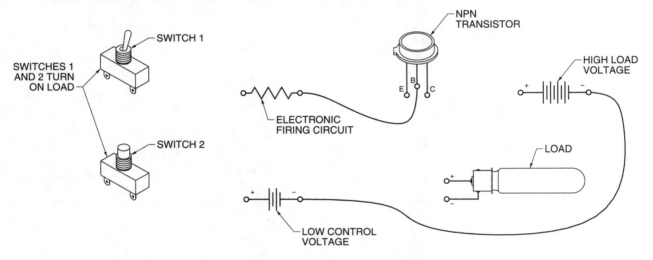

2. Connect the NPN transistor in the circuit so that switch 1 or switch 2 controls the current flow to the transistor, the transistor controls the current flow to the solid-state relay, and the solid-state relay controls the current flow to the load.

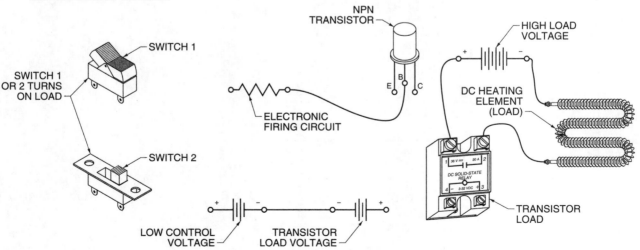

Name _____ Date _____

PNP Transistor Connections

1. Connect the PNP transistor in the circuit so that switch 1 and switch 2 controls the current flow to the transistor and the transistor controls the current flow to the load.

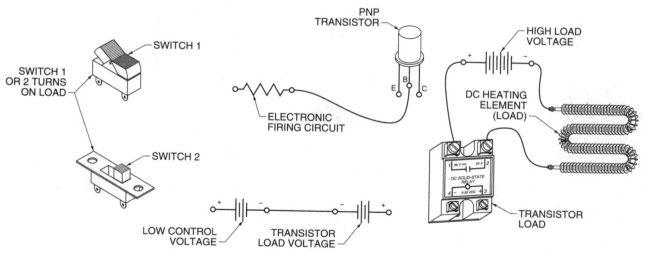

2. Connect the PNP transistor in the circuit so that switch 1 or switch 2 controls the current flow to the transistor, the transistor controls the current flow to the solid-state relay, and the solid-state relay controls the current flow to the load.

Name _____ **Date** _____

SCR Connections

1. Connect the SCR in the circuit so that switch 1 and switch 2 controls the current flow to the SCR, the SCR controls the current flow to the solid-state relay, and the solid-state relay controls the current flow to the load.

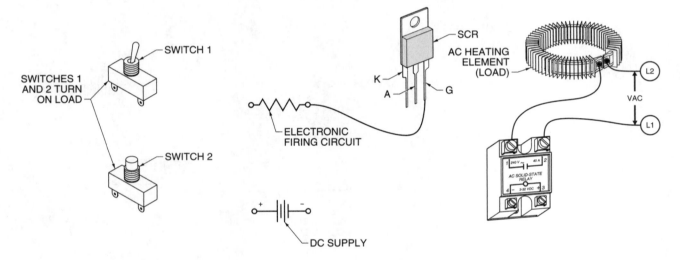

2. Connect the SCR in the circuit so that switch 1 or switch 2 controls the current flow to the SCR and the SCR controls the current flow to the load.

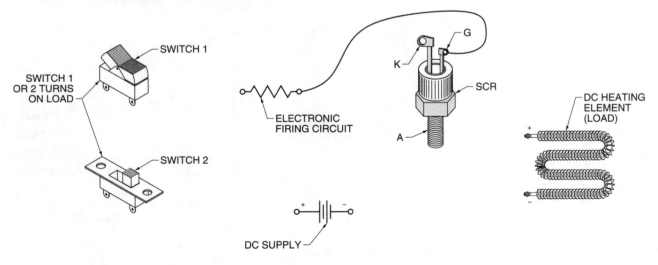

Name _____ Date _____

Triac Connections

1. Connect the triac in the circuit so that switch 1 and switch 2 controls the current flow to the triac and the triac controls the current flow to the load.

2. Connect the triac in the circuit so that switch 1 or switch 2 controls the current flow to the triac and the triac controls the current flow to the load.

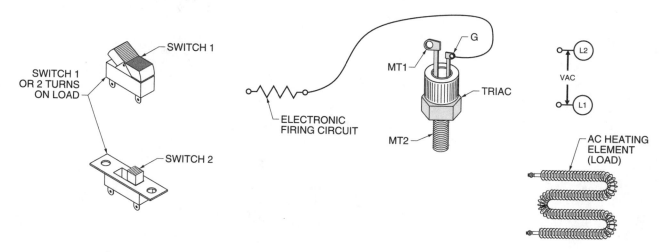

Name _____ Date _____

Electronic Device Connections

1. Complete the wiring diagram based on the schematic diagram.

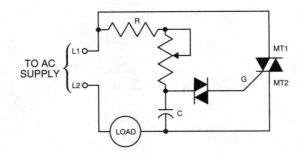

SCHEMATIC DIAGRAM

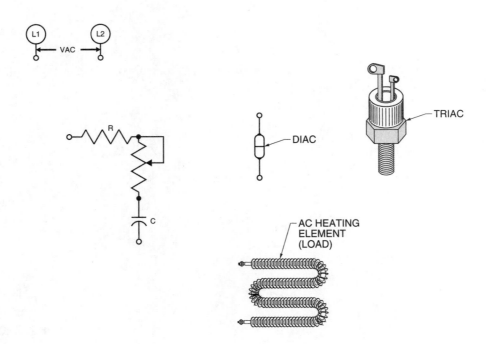

Electronic Control Devices

Activities

24

Name _____ **Date** _____

Hands-On Activity Objectives

- Install a diode in a circuit and determine if the diode is forward or reverse biased.
- Draw the correct diode symbol and direction based on whether the diode is forward or reverse biased.
- Build and test a security system that sounds an alarm if a door or window is opened or a wire is cut.
- Add lamps into a circuit to indicate the operating condition of a security system.
- Add diodes into a circuit so that a back-up power supply can be added into the circuit.

Procedure 24-1

Diodes are used in electronic circuits to allow current to flow in only one direction. Forward bias is the condition of a diode when it allows current flow. Reverse bias is the condition of a diode when it does not allow current flow. Build Circuit 1 inserting the diode in the blank space. The orientation of the diode is arbitrary.

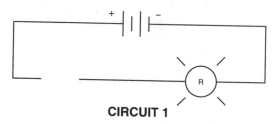

CIRCUIT 1

_____ **1.** Based on the observation of the lamp, is the diode forward or reverse biased?

2. Based on the observation in Step 1, draw the correct position of the diode symbol in Circuit 1.

Build Circuit 2 by inserting the diode in the blank space in the opposite orientation of Circuit 1.

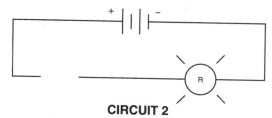

CIRCUIT 2

_____ **3.** Based on the observation of the lamp, is the diode forward or reverse biased?

4. Based on the observation in Step 3, draw the correct position of the diode symbol in Circuit 2.

Procedure 24-2

Security systems must be designed and installed so that they are tamper resistant and should include a back-up power supply in case the primary power is removed.

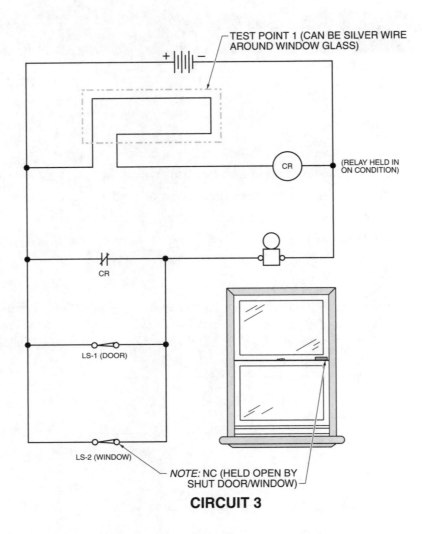

CIRCUIT 3

Build and test Circuit 3. Place an object on LS-1 to simulate a closed door or window and on LS-2 to simulate the closed cover of the security system.

1. Does the alarm sound if LS-1 and/or LS-2 are returned to their normal (nonactivated) positions?

2. What happens if the circuit is opened (cut) at test point 1?

Disconnect the power supply and add Improvement 1 to the circuit. Reconnect the power supply and test the improved circuit.

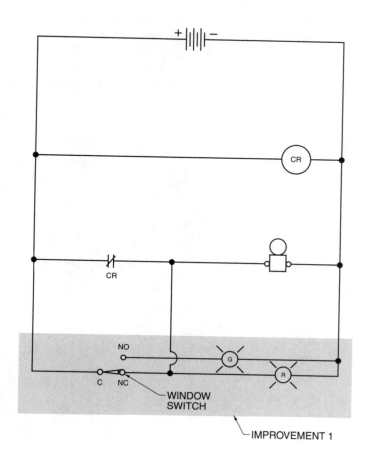

3. When is the red lamp ON?

4. When is the green lamp ON?

Circuit 4 shows the location where a back-up power supply can be added into the circuit so that if the main power is lost or disconnected, the alarm will still sound. In such a circuit, a diode should be added to prevent the voltage of the back-up power supply from flowing into the main power supply but still allow the alarm to sound when the normally closed contacts of the relay are closed. Likewise, a second diode should be added to prevent the voltage of the main power supply from flowing into the back-up power supply. The addition of a second diode also allows the back-up power supply to be at a different voltage than the main power supply. For example, the main power supply could be a 120 VAC to 12 VDC power supply that is plugged into a standard receptacle and the back-up power supply could be a 9 VDC battery.

5. Add two diodes into Circuit 4 (with proper biasing) so that no power from the main power supply can backfeed into the back-up power supply.

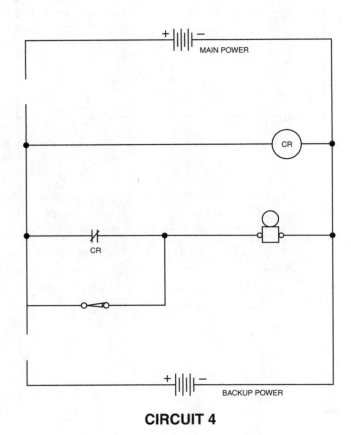

CIRCUIT 4

Name _____ Date _____

True-False

T   F   **1.** A truth table is a diagram that shows graphically how the output of a circuit changes in response to input changes.

T   F   **2.** An op-amp is a general-purpose, high-gain, DC or AC amplifier.

T   F   **3.** A pull-up resistor is a resistor that has one side connected to ground at all times and the other side connected to the unused gate(s).

T   F   **4.** A binary signal is a signal that has only two states: high (1) or low (0).

Multiple Choice

_____ **1.** A(n) ___ gate is a logic gate that provides logic level 1 if one or more inputs are at logic level 1.
   A. AND
   B. NAND
   C. OR
   D. NOR

_____ **2.** A(n) ___ gate is a logic gate that provides logic level 0 only if all inputs are at logic level 1.
   A. AND
   B. NAND
   C. OR
   D. NOR

_____ **3.** A(n) ___ gate is a logic gate that provides logic level 1 only if all inputs are at logic level 1.
   A. XOR
   B. NAND
   C. OR
   D. AND

_____ 4. A(n) ___ gate is a logic gate that provides logic level 1 only if one, but not both, of the inputs are at logic level 1.
    A. NAND
    B. NOR
    C. XOR
    D. NOT

_____ 5. A(n) ___ gate is a logic gate designed to provide logic level 0 if one or more inputs are at logic level 1.
    A. AND
    B. OR
    C. NOT
    D. NOR

Completion

_____ 1. ___ is the output (display) of circuits with continuously changing quantities that may have any value between the defined limits.

_____ 2. ___ is the ratio of the amplitude of an output signal to the amplitude of an input signal.

_____ 3. A(n) ___ IC is an IC that contains an amplifying integrated circuit that produces an output signal that is proportional to the applied input signal.

_____ 4. A(n) ___ input is a digital input that is not a true high or low at all times.

_____ 5. A(n) ___ resistor is a resistor that has one side connected to the power supply at all times and the other side connected to the unused gate(s).

_____ 6. ___ is the output (display) of only specific quantities that change in discrete steps, such as the seconds, minutes, and hours displayed on a digital watch.

_____ 7. ___ are electronic devices in which all components (transistors, diodes, and resistors) are contained in a single package or chip.

_____ 8. A(n) ___ table is a table used to describe the output condition of a logic gate or combination of gates for every possible input condition.

_____ 9. ___ is the number of loads that can be driven by the output of a logic gate.

_____ 10. A(n) ___ gate is a logic gate designed to provide an output that is the opposite of the input.

Digital Electronic Circuits

Worksheet 25-1

25

Name _____ **Date** _____

Truth Tables

Complete the truth tables by using a "0" to indicate when each lamp is OFF and a "1" to indicate when each lamp is ON for each output (L1, L2, and L3) based on the input combination.

1.

INPUTS				OUTPUTS	
PB1	PB2	PB3	PB4	L1	L2
0	0	0	0		
1	0	0	0		
0	0	1	0		
1	1	1	1		

INPUTS	OUTPUTS
1 = ⊸⎯⊸	1 = LAMP ON
0 = ⊸ ⊸	0 = LAMP OFF

2.

INPUTS					OUTPUTS		
PB1	PB2	PB3	PB4	PB5	L1	L2	L3
0	0	0	0	0			
1	1	0	0	0			
1	1	1	0	0			
1	0	1	0	1			
0	0	1	1	0			
0	1	1	1	0			
1	1	1	1	0			
1	1	1	1	1			

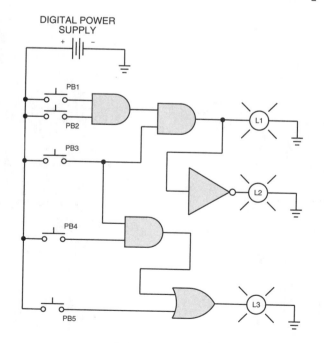

Name _____ Date _____

Electrical to Digital Circuit Conversion

Convert each electrical circuit to its equivalent digital circuit.

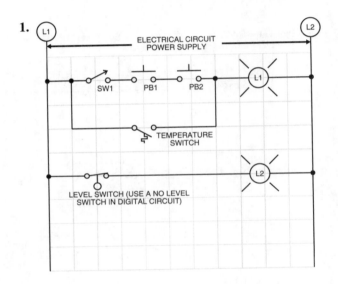

DIGITAL POWER
SUPPLY

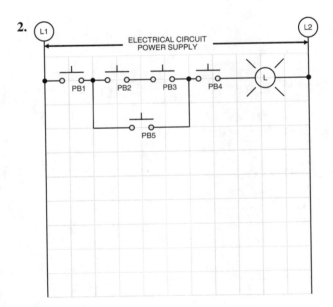

Name _____ Date _____

Digital to Electrical Circuit Conversion

Convert each digital circuit to its equivalent electrical circuit.

1.

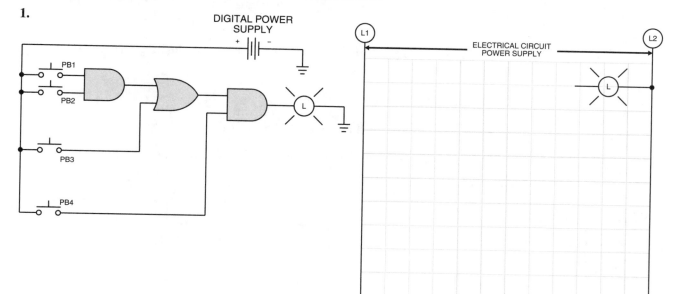

2.

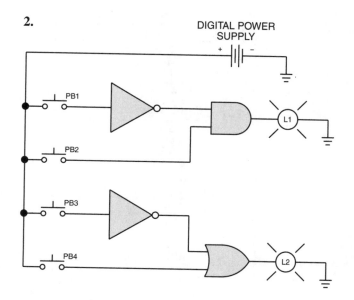

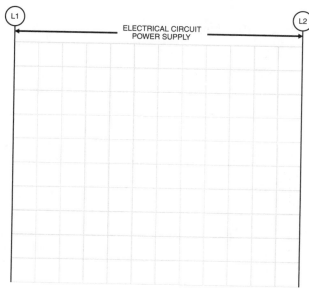

Digital Electronic Circuits

Worksheet 25-4

Name _____ Date _____

Digital Circuit Application

_____ 1. The blue lamp turns ____ if the system temperature falls below the set limit (switch open).

_____ 2. The yellow lamp turns ____ if either one or both of the limit switches detecting the position of the guards are open.

_____ 3. The white lamp turns ON when ____.

_____ 4. ____ switch(es) must be closed to turn ON the red lamp.

_____ 5. ____ switch(es) must be closed to turn ON the green lamp.

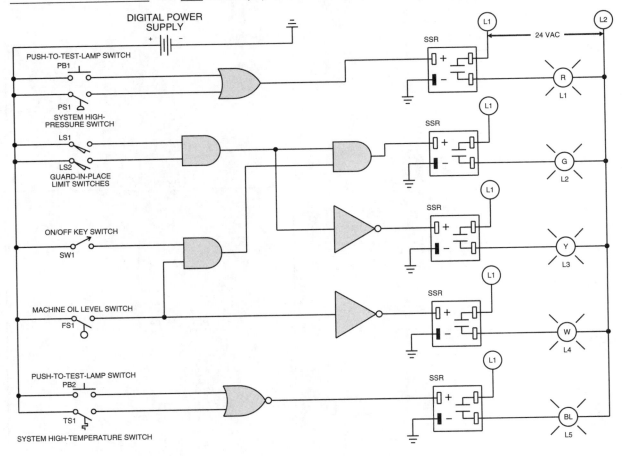

Name _____ Date _____

Truth Table Application

Complete the truth table by using a "0" to indicate when each lamp is OFF and a "1" to indicate when each lamp is ON for each output (L1, L2, and L3) based on the input combination.

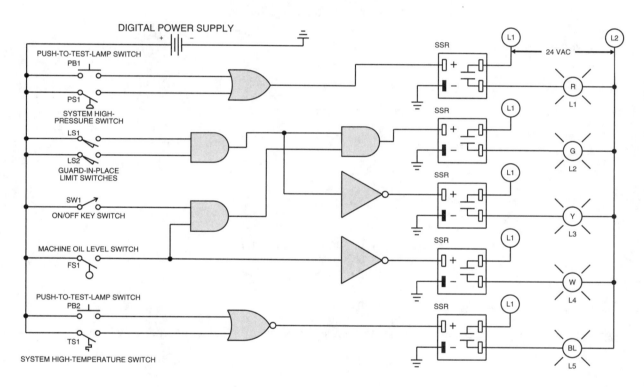

1.

INPUTS								OUTPUTS				
PB1	PS1	LS1	LS2	S1	FS1	PB2	TS1	L1	L2	L3	L4	L5
0	0	0	0	0	0	0	0					
1	0	0	1	1	0	0	0					
0	1	1	0	1	1	0	0					
0	0	0	0	1	1	1	0					
0	0	0	1	1	1	0	1					
0	0	1	1	1	1	1	1					
0	0	1	0	1	1	0	0					
0	0	0	0	1	1	0	0					
1	1	1	1	1	1	1	1					

Name _____ Date _____

Common Control Circuits

_____ **1.** Pushbutton ___ is the motor start pushbutton.

_____ **2.** Pushbutton ___ is the motor stop pushbutton.

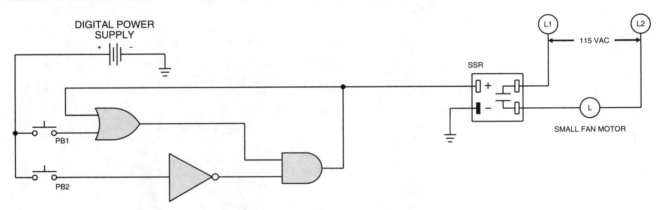

_____ **3.** Pushbutton ___ is the motor start pushbutton.

_____ **4.** Pushbutton ___ is the motor stop pushbutton.

_____ **5.** Pushbutton ___ is the motor jog pushbutton.

_____ **6.** Could the fan motor and solenoid ever be ON simultaneously?

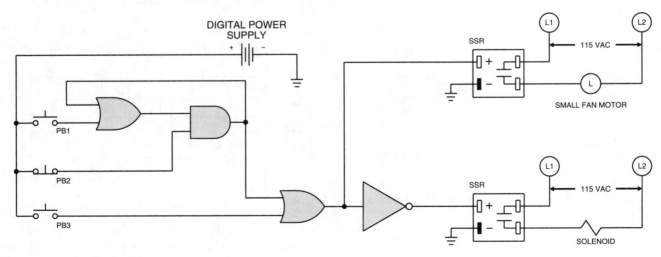

Name _____ Date _____

Programming Activity Objectives

- Program and test a three-input AND logic circuit using simulation software and complete an AND logic truth table based on the observed circuit operation.

- Program and test a three-input OR logic circuit using simulation software and complete an OR logic truth table based on the observed circuit operation.

- Program and test a three-input NAND logic circuit using simulation software and complete a NAND logic truth table based on the observed circuit operation.

- Program and test a three-input NOR logic circuit using simulation software and complete a NOR logic truth table based on the observed circuit operation.

- Program and test a two-input XOR logic circuit using simulation software and complete an XOR logic truth table based on the observed circuit operation.

Procedure 25-1

Program three input switches connected to an AND logic block to control an output device, test the circuit using the TECO SG2 Client simulation software, and complete the truth table.

1. Open the SG2 Client simulation software.

2. Open a new function block diagram document by clicking on the New FBD icon located directly under the Help icon.

3. Open a new circuit by clicking of the New Circuit icon located directly under the File icon at the top of the screen.

4. In the Select Type window, select an SG2-12HR-D model.

5. Place three inputs devices (I-01, I-02, and I-03) on the left side of the screen using the tool palette at the bottom of the screen. Inputs devices are found by opening the Co icon and clicking on I-IN for inputs.

6. Place one output device (Q-01) on the right side of the screen using the tool palette at the bottom of the screen. Outputs devices are found by opening the Co icon and clicking on Q-OUT for outputs.

7. Place one AND logic block between the input devices and output device. Logic blocks are found by opening the LB tool palette at the bottom of the screen.

8. Connect the input devices to the left side of the logic block and the output device to the right side of the logic block using the Connect icon at the bottom of the screen. Connect the lines as required.

9. Select the Simulator icon at the top of the screen (located to the left of the Run icon).

10. Using the Input Status Tool window, test the circuit by clicking on each input device (I-01, I-02, and I-03) to turn each switch ON and OFF. Observe the circuit and status conditions of the inputs and output.

11. Complete the truth table.

12. End the simulation by clicking the Stop icon at the top of the screen.

NOTE:
1 = SWITCH CLOSED
0 = SWITCH OPEN

AND			
INPUTS			OUTPUT
I-01	I-02	I-03	Q-01 (ON or OFF)
0	0	0	
0	0	1	
0	1	0	
1	0	0	
1	0	1	
1	1	1	

13. Print a copy of the circuit by clicking the Print icon at the top of the screen.

14. Save the program as required.

Procedure 25-2

Program three input switches connected to an OR logic block to control an output device, test the circuit using the SG2 Client software, and complete the truth table.

1. Open a new circuit by clicking on the New Circuit icon located directly under the File icon at the top of the screen.

2. Place three inputs devices (I-01, I-02, and I-03) on the left side of the screen using the tool palette at the bottom of the screen.

3. Place one output device (Q-01) on the right side of the screen using the tool palette at the bottom of the screen.

4. Place one OR logic block between the input devices and output device.

5. Connect the inputs and output to the logic block.

6. Select the Simulator icon at the top of the screen.

7. Using the Input Status Tool window, test the circuit by clicking on each input device (I-01, I-02, and I-03) to turn each switch ON and OFF. Observe the circuit and status conditions of the inputs and output.

8. Complete the truth table.

9. End the simulation by clicking the Stop icon at the top of the screen.

NOTE:
1 = SWITCH CLOSED
0 = SWITCH OPEN

OR			
INPUTS			OUTPUT
I-01	I-02	I-03	Q-01 (ON or OFF)
0	0	0	
0	0	1	
0	1	0	
1	0	0	
1	0	1	
1	1	1	

10. Print a copy of the circuit by clicking the Print icon at the top of the screen.

11. Save the program as required.

Procedure 25-3

Program three input switches connected to a NAND logic block to control an output device, test the circuit using the SG2 Client software, and complete the truth table.

1. Open a new circuit by clicking on the New Circuit icon located directly under the File icon at the top of the screen.

2. Place three inputs devices (I-01, I-02, and I-03) on the left side of the screen using the tool palette at the bottom of the screen.

3. Place one output devices (Q-01) on the right side of the screen using the tool palette at the bottom of the screen.

4. Place one NAND logic block between the input devices and output device.

5. Connect the inputs and output to the logic block.

6. Select the Simulator icon at the top of the screen.

7. Using the Input Status Tool window, test the circuit by clicking on each input device (I-01, I-02, and I-03) to turn each switch ON and OFF. Observe the circuit and status conditions of the inputs and output.

8. Complete the truth table.

9. End the simulation by clicking the Stop icon at the top of the screen.

10. Print a copy of the circuit by clicking the Print icon at the top of the screen.

11. Save the program as required.

NAND			
INPUTS			OUTPUT
I-01	I-02	I-03	Q-01 (ON or OFF)
0	0	0	
0	0	1	
0	1	0	
1	0	0	
1	0	1	
1	1	1	

NOTE:
1 = SWITCH CLOSED
0 = SWITCH OPEN

Procedure 25-4

Program three input switches connected to a NOR logic block to control an output device, test the circuit using the SG2 Client software, and complete the truth table.

1. Open a new circuit by clicking on the New Circuit icon located directly under the File icon at the top of the screen.

2. Place three input devices (I-01, I-02, and I-03) on the left side of the screen using the tool palette at the bottom of the screen.

3. Place one output (Q-01) on the right side of the screen using the tool palette at the bottom of the screen.

4. Place one NOR logic block between the input devices and output device.

5. Connect the inputs and output to the logic block.

6. Select the Simulator icon at the top of the screen.

7. Using the Input Status Tool window, test the circuit by clicking on each input device (I-01, I-02, and I-03) to turn each switch ON and OFF. Observe the circuit and Status conditions of the inputs and output.

8. Complete the truth table.

9. End the simulation by clicking the Stop icon at the top of the screen.

10. Print a copy of the circuit by clicking the Print icon at the top of the screen.

11. Save the program as required.

NOTE:
1 = SWITCH CLOSED
0 = SWITCH OPEN

NOR			
INPUTS			OUTPUT
I-01	I-02	I-03	Q-01 (ON or OFF)
0	0	0	
0	0	1	
0	1	0	
1	0	0	
1	0	1	
1	1	1	

Procedure 25-5

Program two input switches connected to an XOR logic block to control an output device, test the circuit using the SG2 Client software, and complete the truth table.

1. Open a new circuit by clicking on the New Circuit icon located directly under the File icon at the top of the screen.

2. Place two inputs devices (I-01 and I-02) on the left side of the screen using the tool palette at the bottom of the screen.

3. Place one output device (Q-01) on the right side of the screen using the tool palette at the bottom of the screen.

4. Place one XOR logic block between the input devices and output device.

5. Connect the inputs and output to the logic block.

6. Select the Simulator icon at the top of the screen.

7. Using the Input Status Tool window, test the circuit by clicking on each input device (I-01 and I-02) to turn each switch ON and OFF. Observe the circuit and Status conditions of the inputs and output.

8. Complete the truth table.

9. End the simulation by clicking the Stop icon at the top of the screen.

10. Print a copy of the circuit by clicking the Print icon at the top of the screen.

11. Save the program as required.

NOTE:
1 = SWITCH CLOSED
0 = SWITCH OPEN

XOR		
INPUTS		OUTPUT
I-01	I-02	Q-01 (ON or OFF)
0	0	
0	1	
1	0	
1	1	

Ohmmeter—Resistance Measurement _____ 273

In-Line Ammeter—DC Measurement _____ 274

Industrial Electrical Symbols _____ 275

Hazardous Locations _____ 279

Chemical Elements _____ 280

Ohmmeter — Resistance Measurement

An ohmmeter measures resistance with all power to the circuit OFF. Low voltage applied to a meter set to measure resistance causes inaccurate readings. High voltage applied to a meter set to measure resistance causes meter damage. Check for voltage using a voltmeter.

Warning: Ensure that no voltage is present in the circuit or component under test before taking resistance measurements.

To measure resistance using an ohmmeter, apply the procedure:

1. Check to ensure that all power is OFF in the circuit or component under test and disconnect the component from the circuit.

2. Set the function switch to the resistance position, which is marked Ω on digital meters.

3. Plug the black test lead into the common jack.

4. Plug the red test lead into the resistance jack.

5. Ensure that the meter batteries are in good condition. The battery symbol is displayed when the batteries are low. Digital meters are zeroed by an internal circuit.

6. Connect the meter test leads across the component under test. Ensure that contact between the test leads and the circuit is good. Dirt, solder flux, oil, and other foreign substances greatly affect resistance readings.

7. Read the resistance displayed on the meter. Check the circuit schematic for parallel paths. Parallel paths with the resistance under test cause reading errors. Do not touch exposed metal parts of the test leads during the test. The resistance of a person's body can cause reading errors.

8. Turn the meter OFF after measurements are taken to save battery life.

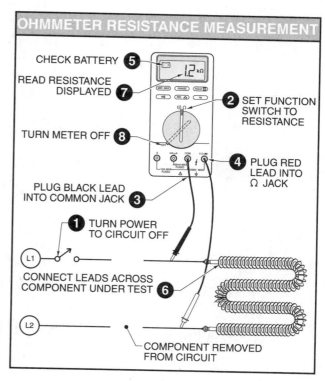

OHMMETER RESISTANCE MEASUREMENT

CHECK BATTERY 5

READ RESISTANCE DISPLAYED 7

1.2 kΩ

SET FUNCTION SWITCH TO RESISTANCE 2

TURN METER OFF 8

PLUG RED LEAD INTO Ω JACK 4

PLUG BLACK LEAD INTO COMMON JACK 3

TURN POWER TO CIRCUIT OFF 1

CONNECT LEADS ACROSS COMPONENT UNDER TEST 6

COMPONENT REMOVED FROM CIRCUIT

In-Line Ammeter — DC Measurement

Care must be taken to protect the meter, circuit, and person using the meter when measuring DC with an in-line ammeter. Always apply the following rules when using an in-line ammeter:

- **Always wear proper protective equipment when working around energized circuits.**

- **Check that the power to the test circuit is OFF before connecting and disconnecting test leads.**

- **Do not change the position of any switches or settings on the meter while the circuit under test is energized.**

- **Turn the power to the meter and circuit OFF before any meter settings are changed.**

- **Connect the ammeter in series with the component(s) to be tested.**

- **Do not take current readings from any circuit in which the current may exceed the limit of the meter.**

Many meters include a fuse in the low-ampere range to prevent meter damage caused by excessive current. Before using a meter, check to see if the meter is fused on the current range being used. The meter is marked as fused or not fused at the test lead terminals. An external fuse may be connected in series with the meter test leads if the meter is not fused. To protect the meter, the fuse rating should not exceed the current range of the meter.

Warning: Ensure that no body parts contact any part of the live circuit, including the metal contact points at the tip of the test leads.

To measure DC using an in-line ammeter, apply the procedure:

1. Set the selector switch to DC. Select a setting high enough to measure the highest possible circuit current if the meter has more than one DC position.

2. Plug the black test lead into the common jack. The common jack may be marked com (common), – (negative), or lo (low).

3. Plug the red test lead into the current jack. The current jack may be marked + (positive), mA (milliamps), or hi (high).

4. Turn the power to the circuit or device under test OFF and discharge all capacitors if possible.

5. Open the circuit and connect the test leads to each side of the opening. The black (negative) test lead is connected to the negative side of the opening and the red (positive) test lead is connected to the positive side of the opening. Reverse the black and red test leads if a negative sign appears in front of the displayed reading.

6. Turn the power to the circuit under test ON.

7. Read the current displayed on the meter.

8. Turn the power OFF and remove the meter from the circuit.

The same procedure is used to measure AC with an in-line ammeter, except that the selector switch is set on AC current.

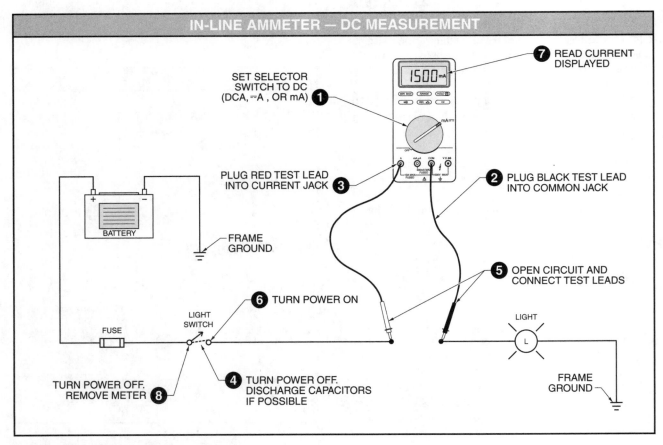

IN-LINE AMMETER — DC MEASUREMENT

INDUSTRIAL ELECTRICAL SYMBOLS . . .

DISCONNECT	CIRCUIT INTERRUPTER	CIRCUIT BREAKER WITH THERMAL OL	CIRCUIT BREAKER WITH MAGNETIC OL	CIRCUIT BREAKER W/ THERMAL AND MAGNETIC OL

LIMIT SWITCHES

NORMALLY OPEN	NORMALLY CLOSED	FOOT SWITCHES	PRESSURE AND VACUUM SWITCHES	LIQUID LEVEL SWITCH	TEMPERATURE-ACTUATED SWITCH	FLOW SWITCH (AIR, WATER, ETC.)

HELD CLOSED | HELD OPEN

NO / NC

SPEED (PLUGGING) | ANTI-PLUG | SYMBOLS FOR STATIC SWITCHING CONTROL DEVICES

STATIC SWITCHING CONTROL IS A METHOD OF SWITCHING ELECTRICAL CIRCUITS WITHOUT USE OF CONTACTS, PRIMARILY BY SOLID-STATE DEVICES. USE SYMBOLS SHOWN IN TABLE AND ENCLOSE THEM IN A DIAMOND.

INPUT COIL   OUTPUT NO   LIMIT SWITCH NO   LIMIT SWITCH NC

SELECTOR

TWO-POSITION

	J	K
A1	X	
A2		X

X = CONTACT CLOSED

THREE-POSITION

	J	K	L
A1	X		
A2			X

X = CONTACT CLOSED

TWO-POSITION SELECTOR PUSHBUTTON

CONTACTS	SELECTOR POSITION			
	A		B	
	BUTTON		BUTTON	
	FREE	DEPRESSED	FREE	DEPRESSED
1-2	X			
3-4		X	X	X

X = CONTACT CLOSED

PUSHBUTTONS

MOMENTARY CONTACT

SINGLE CIRCUIT	DOUBLE CIRCUIT	MUSHROOM HEAD	WOBBLE STICK
NO	NO AND NC		
NC			

MAINTAINED CONTACT

TWO SINGLE CIRCUIT	ONE DOUBLE CIRCUIT

ILLUMINATED

. . . INDUSTRIAL ELECTRICAL SYMBOLS . . .

CONTACTS

INSTANT OPERATING				TIMED CONTACTS - CONTACT ACTION RETARDED AFTER COIL IS:			
WITH BLOWOUT		WITHOUT BLOWOUT		ENERGIZED		DE-ENERGIZED	
NO	NC	NO	NC	NOTC	NCTO	NOTO	NCTC

OVERLOAD RELAYS

THERMAL	MAGNETIC

SUPPLEMENTARY CONTACT SYMBOLS

SPST NO		SPST NC		SPDT		TERMS
SINGLE BREAK	DOUBLE BREAK	SINGLE BREAK	DOUBLE BREAK	SINGLE BREAK	DOUBLE BREAK	SPST SINGLE-POLE, SINGLE-THROW

DPST, 2NO		DPST, 2NC		DPDT	
SINGLE BREAK	DOUBLE BREAK	SINGLE BREAK	DOUBLE BREAK	SINGLE BREAK	DOUBLE BREAK

TERMS:
SPST SINGLE-POLE, SINGLE-THROW
SPDT SINGLE-POLE, DOUBLE-THROW
DPST DOUBLE-POLE, SINGLE-THROW
DPDT DOUBLE-POLE, DOUBLE-THROW
NO NORMALLY OPEN
NC NORMALLY CLOSED

METER (INSTRUMENT)

INDICATE TYPE BY LETTER

TO INDICATE FUNCTION OF METER OR INSTRUMENT, PLACE SPECIFIED LETTER OR LETTERS WITHIN SYMBOL.

AM or A	AMMETER	VA	VOLTMETER
AH	AMPERE HOUR	VAR	VARMETER
µA	MICROAMMETER	VARH	VARHOUR METER
mA	MILLAMMETER	W	WATTMETER
PF	POWER FACTOR	WH	WATTHOUR METER
V	VOLTMETER		

PILOT LIGHTS

INDICATE COLOR BY LETTER

NON PUSH-TO-TEST	PUSH-TO-TEST

INDUCTORS

IRON CORE

AIR CORE

COILS

DUAL-VOLTAGE MAGNET COILS

HIGH-VOLTAGE	LOW-VOLTAGE
LINK	LINKS
1 2 3 4	1 2 3 4

BLOWOUT COIL

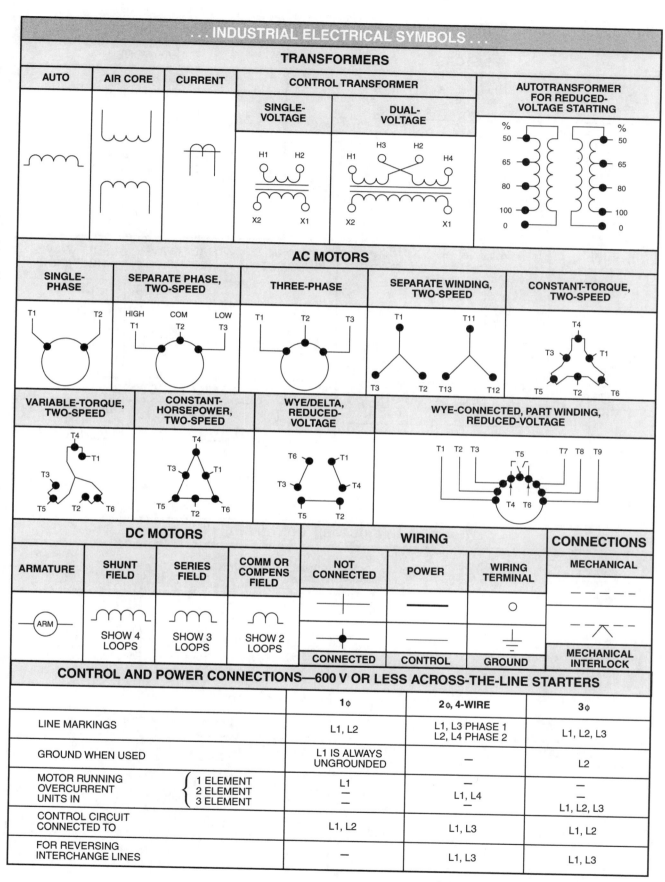

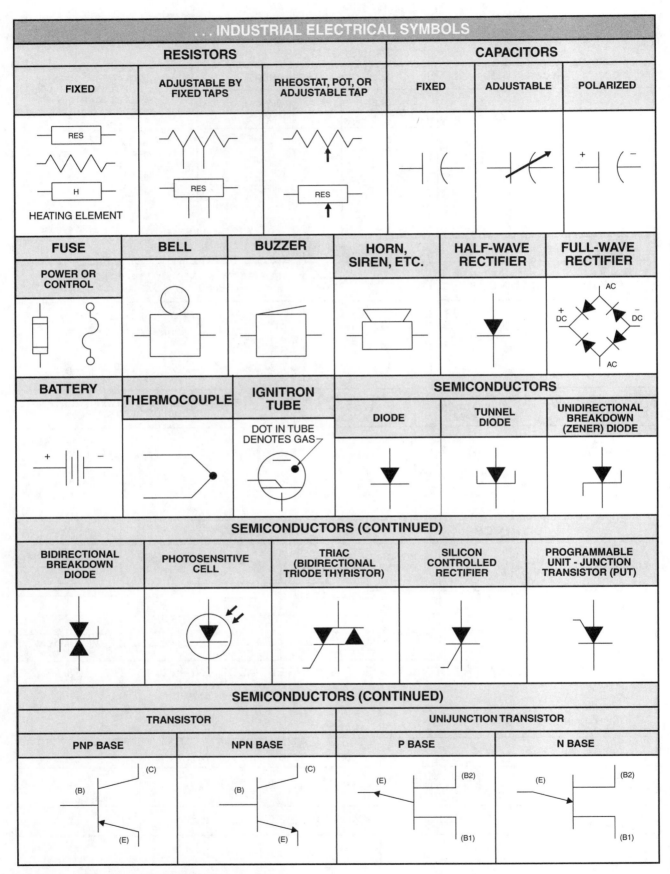

... INDUSTRIAL ELECTRICAL SYMBOLS

Hazardous Locations

Hazardous Location – A location where there is an increased risk of fire or explosion due to the presence of flammable gases, vapors, liquids, combustible dusts, or easily-ignitable fibers or flyings.

Location – A position or site.

Flammable – Capable of being easily ignited and of burning quickly.

Gas – A fluid (such as air) that has no independent shape or volume but tends to expand indefinitely.

Vapor – A substance in the gaseous state as distinguished from the solid or liquid state.

Liquid – A fluid (such as water) that has no independent shape but has a definite volume. A liquid does not expand indefinitely and is only slightly compressible.

Combustible – Capable of burning.

Ignitable – Capable of being set on fire.

Fiber – A thread or piece of material.

Flyings – Small particles of material.

Dust – Fine particles of matter.

DIVISION I EXAMPLES

Class I:

- Spray booth interiors
- Areas adjacent to spraying or painting operations using volatile flammable solvents
- Open tanks or vats of volatile flammable liquids
- Drying or evaporation rooms for flammable vents
- Areas where fats and oil extraction equipment using flammable solvents are operated
- Cleaning and dyeing plant rooms that use flammable liquids that do not contain adequate ventilation
- Refrigeration or freezer interiors that store flammable materials
- All other locations where sufficient ignitable quantities of flammable gases or vapors are likely to occur during routine operations

Classes	Likelihood that a flammable or combustable concentration is present
I	Sufficient quantities of flammable gases and vapors are present in air to cause an explosion or ignite hazardous materials
II	Sufficient quantities of combustible dust are present in air to cause an explosion or ignite hazardous materials
III	Easily ignitable fibers or flyings are present in air, but not in a sufficient quantity to cause an explosion or ignite hazardous materials

Class II:

- Grain and grain products
- Pulverized sugar and cocoa
- Dried egg and milk powders
- Pulverized spices
- Starch and pastes
- Potato and wood flour
- Oil meal from beans and seeds
- Dried hay
- Any other organic material that may produce combustible dusts during their use or handling

Divisions	Location containing hazardous substances
1	Hazardous location in which hazardous substance is normally present in air in sufficient quantities to cause an explosion or ignite hazardous materials
2	Hazardous location in which hazardous substance is not normally present in air in sufficient quantities to cause an explosion or ignite hazardous materials

Class III:

- Portions of rayon, cotton, or other textile mills
- Manufacturing and processing plants for combustible fibers, cotton gins, and cotton seed mills
- Flax processing plants
- Clothing manufacturing plants
- Woodworking plants
- Other establishments involving similar hazardous processes
- or conditions

Groups	Atmosphere containing flammable gases or vapors or combustible dust	
Class I	**Class II**	**Class III**
A	E	none
B	F	
C	G	
D		

Chemical Elements

Name	Symbol	Valence Electrons	Atomic Weight*	Atomic Number	Name	Symbol	Valence Electrons	Atomic Weight*	Atomic Number
Actinium	Ac	2	[227]	89	Neon	Ne	8	20.183	10
Aluminum	Al	3	26.9815	13	Neptunium	Np	2	[237]	93
Americium	Am	2	[243]	95	Nickel	Ni	2	58.71	28
Antimony	Sb	5	121.75	51	Niobium	Nb	1	92.906	41
Argon	Ar	8	39.948	18	Nitrogen	N	5	14.0067	7
Arsenic	As	5	74.9216	33	Nobelium	No	2	[255]	102
Astatine	At	7	[210]	85	Osmium	Os	2	190.2	76
Barium	Ba	2	137.34	56	Oxygen	O	6	15.9994	8
Berkelium	Bk	2	[247]	97	Palladium	Pd	—	106.4	46
Beryllium	Be	2	9.0122	4	Phosphorus	P	5	30.9738	15
Bismuth	Bi	5	208.980	83	Platinum	Pt	1	195.09	78
Boron	B	3	10.811	5	Plutonium	Pu	2	[244]	94
Bromine	Br	7	79.909	35	Polonium	Po	6	[210]	84
Cadmium	Cd	2	112.40	48	Potassium	K	1	39.102	19
Calcium	Ca	2	40.08	20	Praseodymium	Pr	2	140.907	59
Californium	Cf	2	[251]	98	Promethium	Pm	2	[145]	61
Carbon	C	4	12.01115	6	Protactinium	Pa	2	[231]	91
Cerium	Ce	2	140.12	58	Radium	Ra	2	[226]	88
Cesium	Cs	1	132.905	55	Radon	Rn	8	[222]	86
Chlorine	Cl	7	35.453	17	Rhenium	Re	2	186.2	75
Chromium	Cr	1	51.996	24	Rhodium	Rh	1	102.905	45
Cobalt	Co	2	58.9332	27	Roentgenium	Rg	1	[272]	111
Copernicium	Cn	2	[277]	112	Rubidium	Rb	1	85.47	37
Copper	Cu	1	63.54	29	Ruthenium	Ru	1	101.07	44
Curium	Cm	2	[247]	96	Samarium	Sm	2	150.35	62
Darmstadtium	Ds	1	[281.16]	110	Scandium	Sc	2	44.956	21
Dysprosium	Dy	2	162.50	66	Selenium	Se	6	78.96	34
Einsteinium	Es	2	[254]	99	Silicon	Si	4	28.086	14
Erbium	Er	2	167.26	68	Silver	Ag	1	107.870	47
Europium	Eu	2	151.96	63	Sodium	Na	1	22.9898	11
Fermium	Fm	2	[257]	100	Strontium	Sr	2	87.62	38
Flerovium	Uuq	4	[289]	114	Sulfur	S	6	32.064	16
Fluorine	F	7	18.9984	9	Tantalum	Ta	2	180.948	73
Francium	Fr	1	[223]	87	Technetium	Tc	2	[97]	43
Gadolinium	Gd	2	157.25	64	Tellurium	Te	6	127.60	52
Gallium	Ga	3	69.72	31	Terbium	Tb	2	158.924	65
Germanium	Ge	4	72.59	32	Thallium	Tl	3	204.37	81
Gold	Au	1	196.967	79	Thorium	Th	2	232.038	90
Hafnium	Hf	2	178.49	72	Thulium	Tm	2	168.934	69
Helium	He	2	4.0026	2	Tin	Sn	4	118.69	50
Holmium	Ho	2	164.930	67	Titanium	Ti	2	47.90	22
Hydrogen	H	1	1.00797	1	Tungsten	W	2	183.85	74
Indium	In	3	114.82	49	Unnilennium	Une	2	[266]	109
Iodine	I	7	126.9044	53	Unnilhexium	Unh	2	[263]	106
Iridium	Ir	2	192.2	77	Unniloctium	Uno	—	[265]	108
Iron	Fe	2	55.847	26	Unnilpentium	Unp	2	[262]	105
Krypton	Kr	8	83.80	36	Unnilquadium	Unq	2	[261]	104
Lanthanum	La	2	138.91	57	Unnilseptium	Uns	2	[262]	107
Lawrencium	Lr	2	[256]	103	Ununoctium	Uuo	8	[294]	118
Lead	Pb	4	207.19	82	Ununpentium	Uup	5	[288]	115
Lithium	Li	1	6.939	3	Ununseptium	Uus	7	[294]	117
Livermorium	Uuh	6	[293]	116	Ununtrium	Uut	3	[284]	113
Lutetium	Lu	2	174.97	71	Uranium	U	2	238.03	92
Magnesium	Mg	2	24.312	12	Vanadium	V	2	50.942	23
Manganese	Mn	2	54.9380	25	Xenon	Xe	8	131.30	54
Mendelevium	Md	2	[258]	101	Ytterbium	Yb	2	173.04	70
Mercury	Hg	2	200.59	80	Yttrium	Y	2	88.905	39
Molybdenum	Mo	1	95.94	42	Zinc	Zn	2	65.37	30
Neodymium	Nd	2	144.24	60	Zirconium	Zr	2	91.22	40

* a number in brackets indicates the mass number of the most stable isotope